AF358741

Mesoporous Metal Oxide Films

Mesoporous Metal Oxide Films

Editor

Emmanuel Topoglidis

MDPI • Basel • Beijing • Wuhan • Barcelona • Belgrade • Manchester • Tokyo • Cluj • Tianjin

Editor
Emmanuel Topoglidis
University of Patras
Greece

Editorial Office
MDPI
St. Alban-Anlage 66
4052 Basel, Switzerland

This is a reprint of articles from the Special Issue published online in the open access journal *Coatings* (ISSN 2079-6412) (available at: https://www.mdpi.com/journal/coatings/special_issues/metal_oxide_films).

For citation purposes, cite each article independently as indicated on the article page online and as indicated below:

LastName, A.A.; LastName, B.B.; LastName, C.C. Article Title. *Journal Name* **Year**, *Article Number*, Page Range.

ISBN 978-3-03936-884-6 (Hbk)
ISBN 978-3-03936-885-3 (PDF)

Cover image courtesy of Emmanuel Topoglidis.

Contents

About the Editor

Emmanuel Topoglidis received his BSc in Biochemistry from King's College London in 1996, his MSc in Biochemical Research from Imperial College in 1997, and his Ph.D. in Biosensors based on Metal Oxide Films from Imperial College in 2001. His research interests and expertise lie in the areas of electrochemical biosensors, nanomaterials, and nanotechnology. He has worked as a research fellow at the Center of Electronic Materials and Devices of Imperial College and as manager of European and National programs at the MESL laboratory of the NCSR, Demokritos in Athens. He was appointed CTO of the U.S.-based diagnostics company Acron Genomics Inc. in 2006 and Director of the U.K. based medical diagnostics company, Molecular Vision in 2008. In 2012, he was appointed Lecturer, and on April 2019 Assistant Professor, at the department of Materials Science of University of Patras in Greece. He is a member of the International Society of Electrochemistry (ISE) and a member of the Royal Society of Chemistry (RSC). He ranks among the first to extensively research the use of metal oxide films for biosensing applications. His work has been cited over 1500 times worldwide. He has presented his work in more than 35 national and international conferences. He is a reviewer in 25 international peer-reviewed journals.

Editorial

Mesoporous Metal Oxide Films

Emmanuel Topoglidis

Department of Materials Science, University of Patras, 26504 Patras, Greece; etop@upatras.gr;
Tel.: +30-2610-996322

Received: 6 July 2020; Accepted: 10 July 2020; Published: 13 July 2020

Abstract: Great progress has been made in the preparation and application of mesoporous metal oxide films and materials during the last three decades. Numerous preparation methods and applications of these novel and interesting materials have been reported, and it was demonstrated that mesoporosity has a direct impact on the properties and potential applications of such materials. This Special Issue of Coatings contains a series of ten research articles demonstrating emphatically that various metal oxide materials could be prepared using a number of different methods, and focuses on many areas where these mesoporous materials could be used, such as sensors, solar cells, supercapacitors, photoelectrodes, anti-corrosion agents and bioceramics. Our aim is to present important developments in this fast-moving field, from various groups around the world.

Keywords: metal oxide; mesoporous; sol-gel; sensor; supercapacitor; photoelectrode; corrosion; dye sensitized solar cell; PEG; bioceramics; TiO_2; SnO_2; ZnO; NiO; Mn_2O_3

1. Introduction

Porous materials have been widely investigated and applied in many fields owing to their outstanding structural properties. According to the definition of International Union of Pure and Applied Chemistry (IUPAC), porous materials can be categorized into three types: microporous materials (pore size < 2 nm), mesoporous materials (2–50 nm) and macroporous materials (pore size > 50 nm). In the past two decades, great progress has been made in the aspects of fabrication and application of mesoporous metal oxides [1–3].

Mesoporous metal oxide films exhibit excellent physicochemical properties, such as large band gap, large surface area, controllable pore size and morphology, good thermal and chemical stabilities, unique optical and electrical properties, non-toxicity and low costs. Many methods of generating mesoporous films have been developed since mesoporous TiO_2 films were synthesized by O'Regan et al. in 1991 [4]. Most notably, the methods that have been used for the preparation of such metal oxide films include sol-gel screen printing, dip coating, spin coating, sputtering spray pyrolysis, atomic layer deposition, electrodeposition and anodic oxidation. A great deal of effort has been made to simplify these methods in order to prepare films faster and in a more reproducible way. Over the last 30 years, films have attracted significant attention for various applications, ranging from dye-sensitized solar cells, adsorption and separation, chemical and biochemical sensors, gas sensors, drug delivery, electrochromic windows, photo and/or electrocatalysis and energy storage devices such as rechargeable batteries and electrochemical supercapacitors [1–6]. More recently, these materials have been used as the scaffold for the development of perovskite solar cells and sensors [7].

Compared with non-porous metal oxides, the most prominent feature is their ability to interact with molecules not only on their outer surface but also on the large internal surfaces of the material, providing more accessible active sites for reactants. These film electrodes with open, interconnected structures ensure the accessibility of reactants to the active surface sites of electrodes by increasing the mass transport. However, their preparation could be a lengthy process that is difficult to accurately reproduce (thickness and uniformity), involving sol-gel synthesis and sintering. Great effort has been made by researchers worldwide to prepare films faster, in a more reproducible way and at lower temperatures to a degree that would allow their commercial application. The aim of this Special Issue was to put together research articles showing recent developments in the preparation and use of mesoporous metal oxide materials.

2. This Special Issue

This Special Issue, entitled "Mesoporous Metal Oxide Films" contains a collection of ten research articles covering fundamental studies and applications of different metal oxide films. Going into detail, Samourganidis et al. [8] investigated the use of a Metglas ribbon substrate modified with a hemin SnO_2 coating for the development of a sensitive magneto-electrochemical sensor for the determination of H_2O_2. The mesoporous SnO_2 films were prepared at low temperatures, using a simple hydrothermal method that is compatible with the Metglas surface. The Hemin/SnO_2-Metglas sensor displayed good stability, reproducibility and selectivity towards H_2O_2.

A facile hydrothermal process was also used by Son et al. [9] in order this time to synthesize well-crystalline mesoporous Mn_2O_3 materials for the fabrication of a pseudocapacitor. These materials exhibited a high surface area and uniformity of unique mesoporous particle morphology, generating many active sites, a fast-ionic transport and enhanced capacitive properties. Chaudhary et al. [10] also used a hydrothermal approach for the controlled growth of gadolinium oxide (Gd_2O_3) nanoparticles in the presence of ethylene glycol (EG) as a structure-controlling and hydrophilic coating source. The structural, optical, photoluminescence, and sensing properties of the prepared materials, as well as their thermal stability, resistance toward corrosion, and decreased tendency toward photobleaching made Gd_2O_3 a good candidate for the electrochemical sensing of p-nitrophenol and hydrazine using voltammetric and amperometric techniques. The developed sensor exhibited good sensitivity, selectivity, repeatability and recyclability.

Gent et al. [11] prepared multi-layered mesoporous thin TiO_2 films as photoelectrodes using an evaporation-induced self-assembly (EISA) method and layer-by-layer deposition. These films represent suitable host structures for the subsequent electrodeposition of plasmonic gold nanoparticles, exhibiting sufficient UV absorption and electrical conductivity as assured by adjusting film thickness and TiO_2 crystallinity. Enhanced activity was observed with each additional layer of TiO_2. As the surface area was increased, it offered access to more active sites and displayed improved transport properties. These films were tested towards the photoelectrochemical oxidation of water under UV illumination and exhibited good electrochemical and mechanical stability. Katsiaounis et al. [12] prepared mesoporous TiO_2 thin films on fluorine-doped indium tin oxide (FTO) glass substrates using a sol-gel route and the "Dr. Blade" technique, allowing them to directly adsorb Ag plasmonic nanoparticles (AgNPs), capped with polyvinyl pyrrolidone (PVP), on their surface. Voltammetric and spectroelectrochemical techniques were used to characterize the electrochemical behavior of composite films. The electrophotochromism of the Ag-TiO_2 composite is due to oxidation/reduction of the AgNPs that form a thin layer of Ag_2O on the metallic core, forming core/shell nanoparticles. This leads to the fabrication of a simple photonic switch. The phenomenon of the plasmon shift is due to a combination of plasmon shift related to the form and dielectric environment of nanoparticles.

Ammar et al. [13] used ZnO nanoclusters for the detection of chloroform ($CHCl_3$) using density functional theory calculations implemented in a Gaussian 09 program. The results revealed that ZnO nanoclusters with a specific geometry and composition are promising candidates for $CHCl_3$-sensing applications. The adsorption of $CHCl_3$ on the oxygenated ZnO reduces its bandgap, and the deposition of O on a ZnO nanocluster increases its sensitivity to $CHCl_3$ and may facilitate $CHCl_3$ removal or detection. Another facile, low-cost hydrothermal method was used by Umar et al. [14] for the synthesis of Fe-doped TiO_2 nanoparticles. These nanoparticles were used to prepare modified glassy carbon electrodes (GCE) for the development of a hydrazine sensor. The electrochemical sensor based on the metal oxide nanoparticles displayed good sensitivity, linear dynamic range, a low limit of detection and is of low cost. Furthermore, the Fe-doped TiO_2-modified GCE showed a negligible interference behavior towards other analytes that could act as interferents on the hydrazine-sensing performance. Yet again, these metal oxide materials are very promising to be used as coatings for the development of sensors.

One of the most frequent uses of mesoporous metal oxide films is their use as working electrodes for the fabrication of dye-sensitized solar cells (DSSC). In this Special Issue, Tsai et al. [15] added NiO nanoparticles to a TiO_2 paste and used a screen-printing method to make a composite film that could be used for the fabrication of a DSSC. The results showed that the addition of NiO nanoparticles to the TiO_2 working electrode inhibited electron transport and prevented electron recombination with the electrolyte. The electron diffusion coefficient decreased following an increase in the amount of NiO added, confirming that NiO inhibited electron transport. The energy level difference between TiO_2 and NiO generated a potential barrier that prevented the recombination of the electrons in the TiO_2 conduction band with the I_3^- of the electrolyte used. Finally, there was an optimal TiO_2–NiO ratio (99:1) in the electrode for increasing the DSSC device efficiency and electron transport.

Singh et al. [16] synthesized guar gum-grafted 2-acrylamido-2-methylpropanesulfonic acid (GG-AMPS) using guar gum and AMPS as the base ingredients. This material was used as a coating on copper to examine its ability to inhibit copper corrosion. Several heteroatoms present in the GC-AMPS coating promote its good binding to the copper surface, thereby reducing the corrosion rate. The weight loss studies revealed good performance of GG-AMPS at a 600 mg/L concentration. The efficiency decreased with the rise in temperature and at higher concentrations of acidic media. However, the efficiency of the inhibition increased with the additional immersion time. Scanning electron microscopy (SEM) and atomic force microscopy (AFM) studies suggested the potential corrosion mitigation of GC-AMPS coatings on copper surfaces in 3.5% NaCl solution. Finally, Kumar et al. [17] extensively studied the effect of polyethylene glycol (PEG) on $Ca_5(PO_4)_2SiO_4$ (CPS) bioceramics. Sol-gel technology was used to produce bioactive and more reactive bioceramic materials. The addition of 5% and 10% PEG significantly affected the porosity and bioactivity of sol-gel-derived CPS and improved its morphology and physiology. The porous structure of CPS revealed that an apatite layer could be generated on its surface when immersed in synthetic body fluid (SBF). The bioactive nature of CPS could make it a suitable material for hard-tissue engineering applications and for drug loading.

In summary, this Special Issue of Coatings comprises a series of research articles demonstrating the potential use of mesoporous metal oxide films and coatings with different morphology and structures in many technological applications, particularly sensors, supercapacitors and solar cells.

Acknowledgments: I would like to thank all the authors for their contributions to this Special Issue of Coatings, reviewers for their constructive comments and editors for their hard work and quick responses.

Conflicts of Interest: The author declares no conflict of interest.

References

1. Ren, Y.; Ma, Z.; Bruce, P.G. Ordered mesoporous metal oxides: Synthesis and applications. *Chem. Soc. Rev.* **2012**, *41*, 4909–4927. [CrossRef] [PubMed]
2. Mierzwa, M.; Lamouroux, E.; Walcarius, A.; Etienne, M. Porous and transparent metal-oxide electrodes: Preparation methods and electroanalytical application prospects. *Electroanalysis* **2018**, *30*, 1–19. [CrossRef]
3. Innocenzi, P.; Malfatti, L. Mesoporous thin films: Properties and applications. *Chem. Soc. Rev.* **2013**, *42*, 4198–4216. [CrossRef] [PubMed]
4. Walcarius, A. Mesoporous materials and electrochemistry. *Chem. Soc. Rev.* **2013**, *42*, 4098–4140. [CrossRef] [PubMed]
5. Orlandi, M.O. *Tin Oxide Materials: Synthesis, Properties and Applications*; Elsevier: Amsterdam, The Netherlands, 2020; pp. 219–246.
6. O'Regan, B.; Gratzel, M. A Low-cost, high-efficiency solar cell based on dye-sensitized colloidal TiO_2 films. *Nature* **1991**, *353*, 737–740. [CrossRef]

7. Nikolaou, P.; Vassilakopoulou, A.; Papadatos, D.; Koutselas, I.; Topoglidis, E. Chemical sensor for CBr_4 based on quasi-2D and 3D hybrid organic-inorganic perovskites immobilized on TiO_2 films. *Mater. Chem. Front.* **2018**, *2*, 730–740. [CrossRef]

8. Samourgkanidis, G.; Nikolaou, P.; Gkovosdis-Louvaris, A.; Sakellis, E.; Blana, I.M.; Topoglidis, E. Simultaneous electrochemical and magnetoelastic sensing of H_2O_2. *Coatings* **2018**, *8*, 284. [CrossRef]

9. Son, Y.-H.; Bui, P.T.M.; Lee, H.-R.; Akhtar, M.S.; Shah, D.K.; Yang, O.-B. A rapid synthesis of mesoporous Mn_2O_3 nanoparticles for supercapacitor applications. *Coatings* **2019**, *9*, 631. [CrossRef]

10. Chaudhary, S.; Kumar, S.; Kumar, S.; Chaudhary, G.R.; Mehta, S.K.; Umar, A. Ethylene glycol functionalized gadolinium oxide nanoparticles as a potential electrochemical sensing platform for hydrazine and p-nitrophenol. *Coatings* **2019**, *9*, 633.

11. Gent, E.; Taffa, D.H.; Wark, M. Multi-layered mesoporous TiO_2 thin film: Photoelectrodes with improved activity and stability. *Coatings* **2019**, *9*, 625. [CrossRef]

12. Ammar, H.Y.; Bardan, H.M.; Umar, A.; Fouad, H.; Alothman, O.Y. ZnO nanocrystal-based chloroform detection: Density Functional Theory (DFT) study. *Coatings* **2019**, *9*, 769. [CrossRef]

13. Katsiaounis, S.; Panidi, J.; Koutselas, I.; Topoglidis, E. Fully reversible electrically induced photochromic-like behaviour of Ag: TiO_2 thin films. *Coatings* **2020**, *10*, 130. [CrossRef]

14. Umar, A.; Harraz, F.A.; Ibrahim, A.A.; Almas, T.; Kumar, R.; Al-Assiri, M.S.; Baskoutas, S. Iron-doped titanium dioxide nanoparticles as potential scaffold for hydrazine chemical sensor applications. *Coatings* **2020**, *10*, 182. [CrossRef]

15. Tsai, C.-H.; Lin, C.-M.; Liu, Y.-C. Increasing the efficiency of dye-sensitized solar cells by adding nickel oxide nanoparticles to titanium dioxide working electrodes. *Coatings* **2020**, *10*, 195. [CrossRef]

16. Singh, A.; Liu, M.; Ituen, E.; Lin, Y. Anti-corrosive properties of an effective guar gum grafted 2-acrylamido-2-methylpropanesulfonic acid (GG-AMPS) coating on copper in 3.5% NaCl solution. *Coatings* **2020**, *10*, 241. [CrossRef]

17. Kumar, P.; Saini, M.; Kumar, V.; Dehiya, B.S.; Sindhu, A.; Fouad, H.; Ahmad, N.; Hashem, M. Polyethylene glycol (PEG) modified porous $Ga_5(PO_4)_2SiO_4$ bioceramics: Structural, morphological and bioactivity analysis. *Coatings* **2020**, *10*, 538. [CrossRef]

Article

Hemin-Modified SnO$_2$/Metglas Electrodes for the Simultaneous Electrochemical and Magnetoelastic Sensing of H$_2$O$_2$

Georgios Samourgkanidis [1], Pavlos Nikolaou [2], Andreas Gkovosdis-Louvaris [2], Elias Sakellis [3], Ioanna Maria Blana [2] and Emmanuel Topoglidis [2,*]

[1] Department of Chemical Engineering, University of Patras, Patras 26504, Greece; ZoraSamourganov@hotmail.com

[2] Department of Materials Science, University of Patras, Patras 26504, Greece; pavlosnikolaou5@gmail.com (P.N.); andreaslouvaris@hotmail.com (A.G.-L.); Blanamaria1453@gmail.com (I.M.B.)

[3] Institute of Nanoscience and Nanotechnology, National Center of Scientific Research, Athens 15310, Greece; e.sakellis@inn.demokritos.gr

* Correspondence: etop@upatras.gr; Tel.: +30-2610-969928

Received: 2 July 2018; Accepted: 11 August 2018; Published: 16 August 2018

Abstract: In this work, we present a simple and efficient method for the preparation of hemin-modified SnO$_2$ films on Metglas ribbon substrates for the development of a sensitive magneto-electrochemical sensor for the determination of H$_2$O$_2$. The SnO$_2$ films were prepared at low temperatures, using a simple hydrothermal method, compatible with the Metglas surface. The SnO$_2$ film layer was fully characterized by X-ray Diffraction (XRD), Scanning Electron Microscopy (SEM), photoluminescence (PL) and Fourier Transform-Infrared spectroscopy (FT-IR). The properties of the films enable a high hemin loading to be achieved in a stable and functional way. The Hemin/SnO$_2$-Metglas system was simultaneously used as a working electrode (WE) for cyclic voltammetry (CV) measurements and as a magnetoelastic sensor excited by external coils, which drive it to resonance and interrogate it. The CV scans reveal direct reduction and oxidation of the immobilized hemin, as well as good electrocatalytic response for the reduction of H$_2$O$_2$. In addition, the magnetoelastic resonance (MR) technique allows the detection of any mass change during the electroreduction of H$_2$O$_2$ by the immobilized hemin on the Metglas surface. The experimental results revealed a mass increase on the sensor during the redox reaction, which was calculated to be 767 ng/μM. This behavior was not detected during the control experiment, where only the NaH$_2$PO$_4$ solution was present. The following results also showed a sensitive electrochemical sensor response linearly proportional to the concentration of H$_2$O$_2$ in the range 1×10^{-6}–72×10^{-6} M, with a correlation coefficient of 0.987 and detection limit of 1.6×10^{-7} M. Moreover, the Hemin/SnO$_2$-Metglas displayed a rapid response (30 s) to H$_2$O$_2$ and exhibits good stability, reproducibility and selectivity.

Keywords: SnO$_2$; Metglas; hemin; H$_2$O$_2$; cyclic voltammetry; magnetoelastic resonance; sensor

---

1. Introduction

Hydrogen peroxide (H$_2$O$_2$) is one of the most important intermediate products of several enzyme-catalyzed oxidation reactions, and an essential substance, analyte or mediator in pharmaceutical, food, clinical and environmental analyses [1–5]. This has raised extensive interest over the years for establishing protocols for H$_2$O$_2$ sensing depending on its application.

Numerous analytical methods have been used for the monitoring of H$_2$O$_2$ including fluorescence, spectrometry, chemiluminescence, colorimetric techniques and liquid chromatography [6–11].

However, most of these techniques are too expensive (equipment and reagents), time-consuming, suffer from interferences and often require complex sample pre-treatment and competent operators to perform the analysis and therefore are not applicable for in situ analysis. Therefore, despite the numerous methods for H_2O_2 detection available, it is still of interest to develop a simple and direct method, that would be reliable, free from interferences and of lower cost. In this sense, electroanalytical methods have been used for the sensitive and selective determination of H_2O_2 [12–14].

Electrochemical sensors (voltammetric or amperometric) are among the most popular sensor devices. They usually monitor the change in current at an applied bias, induced by a redox reaction [15]. Furthermore, based on the properties of the material/substrate used, chemiresistive, capacitive and optical devices have been developed particularly for the sensing of hazardous and toxic gases [16–18]. The signal/outcome in electrochemical sensors depends on the rate of mass transfer to the electrode surface. The aim is to minimize the diffusion path of the detectable analyte and therefore the enzyme should have close contact with the substrate (electrode) used as the transducer. Although in some cases it might be possible to achieve direct electron transfer between the immobilized enzyme and electrode, normally, redox centers of enzymes are located deep in the insulated protein matrix, which makes a direct electron transfer unfeasible. The application of such enzymes in biosensors will require complicated immobilization procedures in order to orient the enzyme molecules in the most efficient way on the surface of the electrode and/or the additional use of redox mediators. In addition, the enzyme molecules tend to aggregate and become inactivated after immobilization and some of them present high cost in their preparation and storage and offer insufficient long-term stability [19–22]. To overcome these obstacles, non-enzymatic H_2O_2 electrochemical sensors have been proposed in recent years replacing immobilized enzymes with various nanocomposites with peroxidase-like activity. These include noble metal nanocomposites [19,20,23–25], metal oxide nanostructures [22,26–28] and metalloporphyrins [28–32].

Hemin (iron protoporphyrin IX chloride) is a well-known natural metalloporphyrin and is the active center of the heme-protein family, which includes hemoglobin, myoglobin and peroxidase. It has a porphyrin ring with an electroactive center of Fe^{3+} and, as a result, a couple of quasi-reversible or reversible redox peaks are usually well-defined in cyclic voltammograms (CVs) obtained in aqueous solution, such as phosphate buffer of wide pH range or in some non-aqueous ones such as hexafluorophosphate ionic liquids. Therefore, hemin has been extensively used to study the redox activity of heme-proteins. In addition, it exhibits remarkable good electrocatalysis towards some small molecules, such as O_2 [33], NO [34], NO_2^- [35], H_2O_2 [28–32], trichloroacetic acid [36], organohalides [37], phenols [38] and artemisinin [39], many of them are related to biological processes. The hemin-Fe acts as the electroactive mediate in the electrocatalysis process.

Hemin exhibits peroxidase-like activity similar to the enzyme [40] and is found to be superior to noble metal catalysts and nanomaterial enzyme mimics, which often exhibit low stability, high cost and poor reproducibility [41,42]. However, the fact that hemin tends to aggregate (inactive dimmers formation) in aqueous media due to its low solubility hinders its direct use as a redox catalyst. Therefore, various methods and electrodes have been used to overcome these issues and develop stable and sensitive hemin-modified electrodes. Hemin has been successfully adsorbed on many carbon materials, incorporated in carbon paste, entrapped in polymeric matrices, immobilized using dendrimers or cationic surfactants, mixing with metal oxide nanoparticles and incorporated or drop casted on various nanocomposite materials.

Our group recently presented [32] a simple and efficient method for the preparation of hemin-modified mesoporous SnO_2 films on low cost flexible, conducting ITO-PET substrates for the electrochemical sensing of H_2O_2. SnO_2 films can be prepared at low temperatures using a simple hydrothermal method, allowing not only high hemin loading in a stable and functional way, but also the direct reduction and oxidation of the immobilized hemin. In another recent work of ours [43], a nanostructured ZnO layer was synthesized onto a Metglas magnetoelastic ribbon to immobilize hemoglobin (Hb) on it and study the Hb's electrocatalytic behavior towards H_2O_2. Hb oxidation by H_2O_2 was monitored simultaneously by two different techniques: Cyclic Voltammetry (CV) and Magnetoelastic Resonance (MR). The Metglas/ZnO/Hb system was simultaneously used as a working electrode for the CV scans and as a magnetoelastic sensor excited by external coils, which drive it to resonance and interrogate it.

Metal oxides are typically wide-bandgap semiconductors and are, therefore, effectively insulating for applied potentials lying within their band gap. The conductivity of nanocrystalline TiO_2 and ZnO electrodes has been shown to be enhanced by a high density of sub-band gap states; nevertheless, such electrodes are still essentially insulators for potentials more positive than -0.3 and -0.15 V, respectively [44]. Consequently, electrochemical studies of heme proteins immobilized on such electrodes have been limited to electrochemical reduction of such proteins, with the dynamics of this reduction having been largely limited by the limited conductivity of the metal oxide film. Previous reports have indicated that mesoporous SnO_2 films are more conductive than either ZnO or TiO_2 films. This metal oxide exhibits a band gap (330 nm) and an isoelectric point (IEP~5) similar to those of TiO_2. Although flat (not porous) indium- or fluorine-doped SnO_2 films have been used as transparent electrodes for protein immobilization [45], the protein loading on such flat electrodes is; however, limited to a monolayer coverage at least 2 orders of magnitude lower than the monolayer coverage that we demonstrate here as possible on mesoporous electrodes.

In this work, we propose to prepare thin mesoporous SnO_2 films on the Metglas ribbon and use it for the immobilization of hemin and examine its sensitivity towards the electrocatalytic reduction of H_2O_2. SnO_2 films are particularly attractive for immobilization of molecules, exhibit a high surface area, non-toxicity, chemical stability and unique electronic and catalytic behavior [32]. They are similar to the ZnO films we used in the past [43], but more conductive and allowing redox reactions to take place at more moderate potentials at their conduction band edge. In addition, their preparation involves very few steps using a simple, low-cost, low-temperature hydrothermal method, which is much simpler than the sol-gel method involving an autoclave reactor that we used for the preparation of ZnO films in the past. Table 1 displays structural and electrochemical properties of mesoporous films of different metal oxides modified with hemin or heme proteins for studying their electrochemical behavior and/or used for the development of biosensors. Although, structurally, most of these films are similar (size of nanoparticles and thickness) and exhibit a high surface area, multistep, lengthy and not always reproducible procedures involving high-temperature sintering are necessary for their preparation. In addition, as they are semiconductors, during the electrochemical measurements they exhibit an insulating region, mostly at positive biases, that hinders the reduction or oxidation of the adsorbed molecules or limits their electrocatalytic efficiency. They exhibit limited conductivity at low negative potentials and a relatively slow electron transport. According to our studies and those of other groups we have come to the conclusion that the SnO_2 films could be prepared by a simple, fast, low-cost and low-temperature hydrothermal route, exhibiting a high surface area for the immobilization of molecules and biomolecules. In addition, and compared with the other metal oxides, they exhibit a very limited insulating region only at high positive biases and could be successfully used for the development of electrocatalytic sensors.

In addition, hemin is a much smaller molecule than the Hb we used in the past and is not surrounded by a shell of amino acids which that only a part of it to come into direct contact with the surface of the electrode [43]. Therefore, electron transfer between hemin and electrode is expected to be faster and more direct. From our previous studies, we found that the SnO_2 films provide a suitable microenvironment to prevent hemin aggregation and dimerization, therefore maintaining its activity after immobilization [32]. Hemin, dissolved in organic solvent, was drop-coated with a pipette on the SnO_2-Metglas substrates.

Based on the physicochemical properties of our proposed Hemin/SnO_2-Metglas sensor, CV and MR will be used simultaneously in order to study the chemical behavior of immobilized hemin with H_2O_2. While the experimental technique of CV can provide information on the electrochemical behavior of a reaction, the MR of magnetoelastic materials can quantify those reactions due to the high sensitivity to external parameters such as mass load [46,47]. As magnetoelasticity, we define the property of some ferromagnetic materials to convert efficiently the magnetic energy into elastic energy. According to Hernado et al. [48] the parameter associated with the energy transfer between the elastic and magnetic subsystems is known as the magnetoelastic coupling coefficient k ($0 < k < 1$). By far the best-known materials with high magnetoelastic coupling coefficients are metallic glasses [49], which are mainly amorphous alloys of magnetic materials in the shape of ribbons. A free-standing ribbon of metallic glass can be easily induced to vibrate mechanically to one of its resonance modes, to an external AC magnetic field, due to its ferromagnetic nature and magnetoelastic property. Those mechanical vibrations depend upon factors such as the stiffness and the mass of the ribbon, and any change on them will affect the dynamic behavior of the ribbon. Specifically, the appearance of extra mass on a magnetoelastic ribbon will affect its vibration behavior, by reducing its resonance frequencies, and thus it will leave a trace of the mass on the ribbon. Exploiting this behavior, information about the interaction of H_2O_2 with the immobilized hemin can be obtained.

The resulting Hemin/SnO_2-Metglas electrodes exhibit highly efficient electrocatalytic reduction of H_2O_2, with good stability and sensitivity and to our knowledge, this is only the second time that the two methods, CV and MR have been used simultaneously for biodetection and the development of a H_2O_2 sensor.

Table 1. Characteristics and properties of metal oxide films used as electrodes.

Electrode	Preparation	Particle Size (nm)	Thickness (μm)	Insulating Region	Ref.
Hemin/SnO_2-ITO/PET	Low temperature Hydrothermal method	20–70	4	At +0.2 V and more positive biases	[32]
Hemin-ZnO-Metglas	Hydrothermal method/sintering	11–32	1	none	[43]
Hemin/TiO_2-FTO [1]	Hydrolysis/sol-gel/sintering	10–15	2	At −0.3 V and more positive biases	[37]
Hemin/TiO_2-GCE [2] electrode	Flame synthesis technique	10–50	10	none	[28]
Cyt-c [3]/SnO_2/FTO Hb [4]/SnO_2/FTO	Sol-gel/sintering	15–20	4	At +0.2 V and more positive biases	[50]
Cyt-c/TiO_2/FTO Cyt-c/ZnO/FTO	Sol-gel/sintering	10–20	4	At −0.15 V and more positive biases	[44]
Hemin/SnO_2-Metglas	Low temperature Hydrothermal method	20–70	11	none	This work

[1] Fluorine doped tin oxide; [2] Glassy carbon electrode; [3] Cytochrome c; [4] Hemoglobin.

2. Materials and Methods

2.1. Materials

A commercial ribbon of Metglas 2826MB ($Fe_{40}Ni_{38}Mo_4B_{18}$) was purchased from Hitachi Metals Europe GmbH (Düsseldorf, Germany). Tin (IV) Oxide Nanopowder, <100 nm average particle size (BET), t-Butanol anhydrous, ≥99.5%, bovine hemin (≥90%), absolute ethanol, analytical grade acetone and sodium dihydrogen orthophosphate (NaH_2PO_4) were all purchased from Sigma-Aldrich Chemie GmbH (Taufkirchen, Germany). Dimethyl sulfoxide (DMSO) was obtained from Fisher Scientific GmbH (Schwerte, Germany) and was of HPLC grade. H_2O_2 (30% *w/v* solution) was purchased from Lach-Ner (Neratovice, Czech Republic) and was diluted. All aqueous solutions were prepared with deionized water.

2.2. Preparation of Hemin/SnO_2–Metglas Electrodes

Tin oxide powder was homogeneously dispersed in a mixture of t-Butanol and acetonitrile 95:5 (*v/v*) at a concentration of $40\ g{\cdot}L^{-1}$ as previously described by our group [32]. The suspension was then sonicated in a JENCONS-PLS sonicator (Jencons, Bedford, UK), while the mixture solution was immersed in an ice bath to regulate its temperature. After sonication, the solution was semi-opaque and the particles evenly distributed.

A commercial ribbon of Metglas 2826MB ($Fe_{40}Ni_{38}Mo_4B_{18}$) was used as the substrate for the deposition of the SnO_2 films. The ribbon was cleaned thoroughly in acetone for 15 min under sonication and then cut in strips of 2 cm in length. The Metglas strips were then placed in a petri dish with their rough side facing upwards. Adhesive Tape of known thickness was then applied on the surface of the Metglas to define the film layer thickness accordingly, imitating a doctor-blade technique. Masking each Metglas substrate with 3M Scotch Magic tape (3M, Berkshire, UK) type 810, thickness 62.5 μm) controlled the thickness and width of the solution spread area. Afterwards, 20 μL of the above-mentioned SnO_2 solution was deposited on to the Metglas surface and left to air-dry at 37 °C for 20 min until all the solvent was totally evaporated giving as a result a thin SnO_2 film layer. One layer of tape was employed for each SnO_2 film deposition, which provided a film thickness of ~7 μm and size of 1×1 cm^2. The thickness of the films was measured by SEM and the use of a standard profilometer. It appears that t-Butanol reduces the surface tension of the liquid paste to improve its adhesion on the Metglas surface.

Afterwards, a solution of 10 μM hemin in DMSO was prepared and 10 μL was drop-coated with a pipette on the thin SnO_2 film. The use of this concentration of hemin was selected based on a recent study of ours [32] involving the immobilization of hemin on semitransparent SnO_2/ITO-PET films. Using UV-Vis absorption spectroscopy (Shimadzu Europa UV-1800, Duisburg, Germany) we confirmed that the immobilized hemin exhibited its characteristic peaks and therefore remained intact. By using higher concentrations of hemin, aggregation of hemin molecules occurred on the surface of the films, affecting its electrocatalytic behavior towards H_2O_2. The films were then left to air-dry at room temperature for 30 min, until there were no signs of moisture on their surface. Once the hemin was adsorbed, it remained strongly bound to the films. Prior to all measurements, films were rinsed with NaH_2PO_4 to remove possible non-adsorbed hemin.

2.3. Characterization of the Hemin/SnO$_2$-Metglas Film Electrodes

The morphology and thickness of the SnO_2-Metglas film electrodes were determined by field emission scanning electron microscopy (FE-SEM), using a FEI Inspect Microscope (Thermo Fisher Scientific, Ferentino, Italy) operating at a voltage of 25 kV. The specimens (films) were prepared by Au sputtering to increase the conductivity of the samples. Energy dispersive spectroscopy (EDS, Thermo Fisher Scientific, Ferentino, Italy) was also used for the elemental analysis of the SnO_2-Metglas film electrodes. As a further proof of the quality of the SnO_2 films obtained, their photoluminescence (PL) spectra were recorded using a Hitachi F2500 Fluorescence Spectrophotometer (Hitachi, Tokyo, Japan) from 350 to 550 nm at an excitation wavelength of 300 nm.

The crystal structure of the SnO_2 films on the Metglas substrate was studied using X-ray diffraction (XRD). The XRD pattern of the SnO_2 film on Metglas was measured using a Bruker D8 advance X-ray diffractometer (Bruker AXS GmbH, Karlsruhe, Germany) with Cu Kα-radiation from 20° to 80° at a scanning speed of 0.015 °/s. The X-ray tube voltage and current were set at 45 kV and 40 mA, respectively. The diffraction patterns were indexed by comparison with the Joint Committee on Powder Diffraction Standards (JCPDS) file 41-1445 of SnO_2 cassiterite [51].

Fourier transform infrared (FTIR) spectroscopic analysis of the SnO_2 films with or without adsorbed hemin was carried out with a Digilab Excalibur FTS 3000MX spectrometer (BioRad Laboratories GmbH, München, Germany).

2.4. Experimental Setup for MR and CV Measurements

Shown in Figure 1 is the experimental setup that was used for simultaneous MR and CV detection of the interaction of immobilized hemin with H_2O_2. To begin, a homemade coil (N = 30 turns, R = 1.1 Ω, L = 15.2 μH), which is connected to a microcontroller frequency generator (Sentech,

Pittsburgh, PA, USA) (Figure 1b), is wrapped around a 29.5 mm diameter syringe and hung firmly in a structure made out of wood, as shown in Figure 1b. The frequency generator drives an alternative current (AC) through the coil, producing an alternative magnetic field, which in turn induces elastic waves on the magnetoelastic sensor inside the coil. When the frequency of the vibration matches the first longitudinal natural frequency of the sensor, resonance occurs. By using software written in Visual Basic, the resonance frequency value of the sensor is recorded. The piston of the syringe is used to secure the working electrode (WE) (our sensor) of the setup and is allowed to move vertically during the experiment, in order to control the immersion position of the sensor. For the CV measurements, a platinum wire 30 mm in length, which was welded onto an extension copper wire, is set up at the side of the syringe to work as the counter electrode (CE) (Figure 1b). As reference electrode (RE), $Ag/AgCl/KCl_{sat}$ was used, positioned in a holder that was made in order to be able to remove it from the setup and store it after each experiment (Figure 1b).

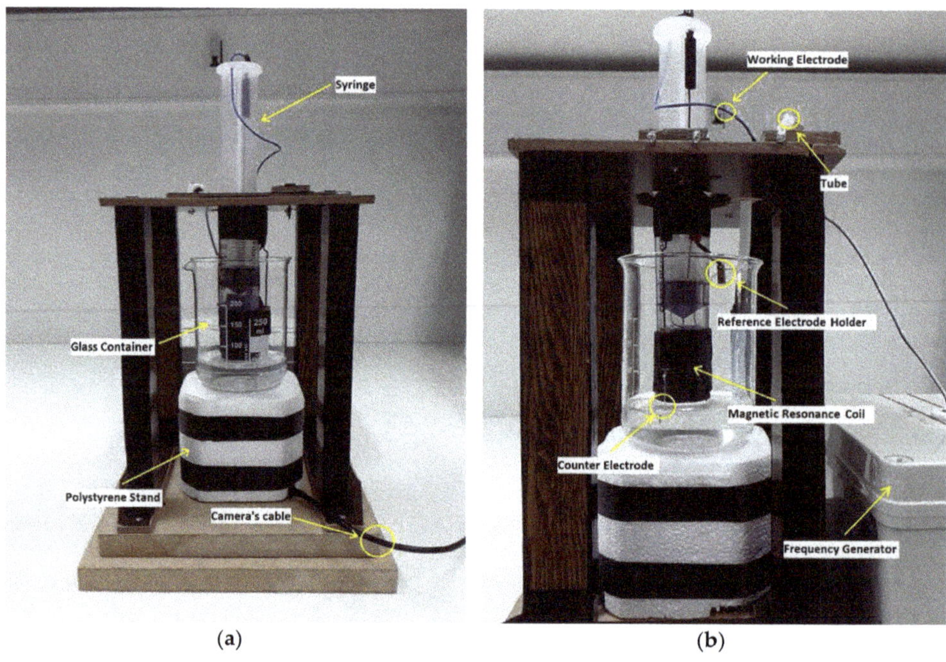

Figure 1. Experimental setup: (**a**) Front view; (**b**) side view.

All three electrodes were immersed into NaH_2PO_4 solution (pH = 7) inside a glass container (Figure 1a), the quantity of which is controlled by an injection tube shown in Figure 1b. A stand made of polystyrene (PS) (Figure 1a) is used to hold the glass container, inside of which a wire camera was wedged (Figure 1a), observing the immersion process of the WE and providing information on whether or not the sensor touched the surface of the solution. Figure 2a shows the WE of the setup with the magnetoelastic sensor being held magnetically on a copper wire with the use of a tiny piece of neodymium magnet welded carefully to the copper wire. In this way, the conductivity of the WE is increased, reducing any noisy effects during CV. Figure 2b,c shows the state before and after the immersion of the WE inside the solution. The six bright spots in the middle of Figure 2b are the reflection of the camera lights at the surface of the solution. The characteristic ellipse of Figure 2c indicates the immersed state of the electrode. It is important to know the exact moment the electrode touches the surface of the solution, because this way, the immersion depth of the electrode can be controlled using the scale on the syringe.

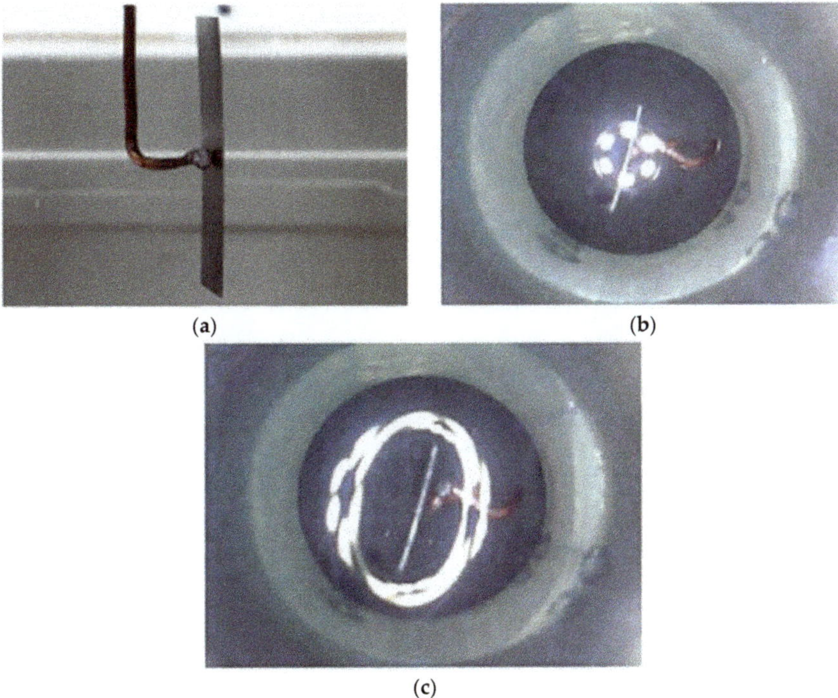

Figure 2. (a) The WE of the setup; (b) WE before and (c) after its immersion into the solution.

2.5. Experimental Preparation and Procedure

The experimental preparation was carried out in two main stages. The first stage concerns the connection of the device, described above, to the other devices used in the experimental procedure. For the MR measurements, a home-made microcontroller frequency generator was connected to a PC through a RS232 serial port, and the acquisition data process was carried out by software written in Visual Basics programming. The wire camera was also connected to the PC and controlled by its software. An Autolab PGStat 101 potentiostat (Metrohm Autolab, Utrecht, The Netherlands) was used for the CV measurements. The second stage includes the preparation and insertion of the Hemin/SnO$_2$-Metglas sensor into the experimental setup. As mentioned above, when the frequency of the vibrated sensor matches with its first longitudinal natural frequency, resonance occurs. The first longitudinal vibration of a beam-like structure with both ends free has its node at the middle and the maximum deformation at the free ends. Figure 3 shows the graphical illustrations of the resonance peaks of a 30 mm bare Metglas ribbon with the copper wire when holding it at different positions (Figure 2a). The first peak (Figure 3a) corresponds to the position at the middle of the sensor, while the other three peaks (Figure 3b–d) correspond to a position 1, 2 and 3 mm from the middle, respectively. As we move the holding point away from the middle, where the node is located, the amplitude of the resonance peak drops rapidly. This happens because the node is a stationary position, and any changes on it do not affect the vibrating behavior of the sensor. On the other hand, changes in the deforming positions affect the behavior of the transmitting elastic waves, which in our situation is the damping effect on the wave due to the physical contact between the sensor and the copper wire. It is crucial to have the amplitude of the resonance peaks as high as possible, before the immersion of the sensor inside the solution, because after the immersion, the damping effect is strong, and in order to not completely lose the peak at the noise level, preparations must be made. For the CV measurements, all applied biases are reported against the Ag/AgCl reference electrode.

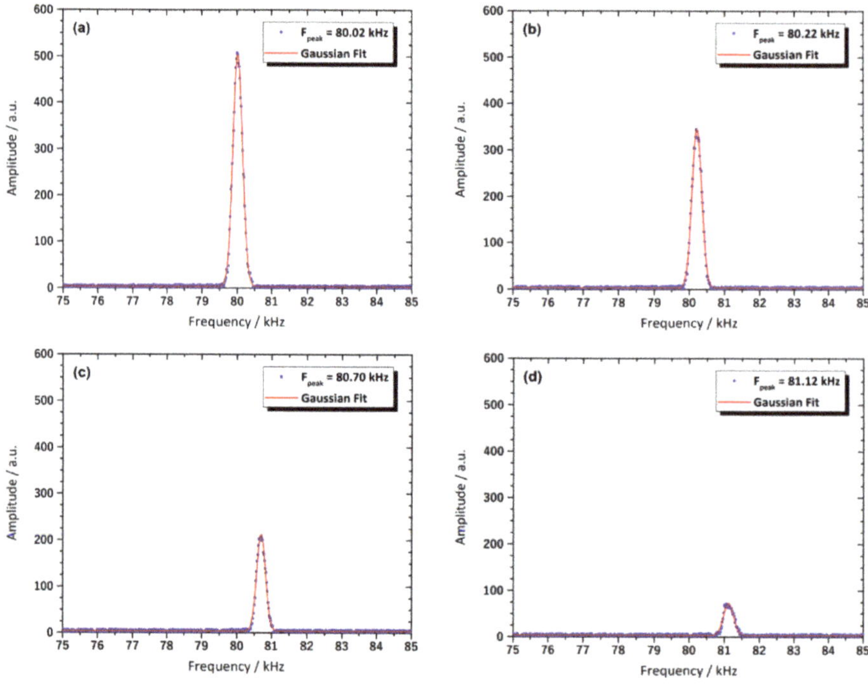

Figure 3. Resonance peaks at different holding positions of the Metglas film: (**a**) Middle; (**b**) 1 mm from the middle; (**c**) 2 mm from the middle; (**d**) 3 mm from the middle.

The electrolyte, an aqueous solution of 10 mM NaH_2PO_4 (pH = 7) was deoxygenated with argon prior to any electrochemical measurements. All experiments were carried out at room temperature.

Upon completion of the preparation and insertion of the $Hemin/SnO_2$-Metglas sensor into the setup, the experimental measurement procedure was carried out. Initially, by moving the syringe piston along the coil and carrying out some MR measurement in the air, the sensor position with the highest resonance peak amplitude was noted using the syringe scale. Next, the immersion process took place in three steps in order to conduct measurements with both CV and MR techniques. In the first step, the piston was moved till the point that the sensor touched the surface of the solution (Figure 2c), using the guidance of the wire camera. In the second step, the piston was moved 4 mm down, using the scale on the syringe, so as to immerse the whole surface of the immobilized hemin into the solution, and a CV measurement was taken. The last step was to return the sensor to its initial position and run a MR measurement. Figure 4 shows the resonance peaks before and after the immersion of a bare Metglas ribbon (Figure 4a,b) and of a $Hemin/SnO_2$-Metglas sensor (Figure 4c,d). In the case of the latter, the decrease on the amplitude and the frequency seems to be higher than that of the bare Metglas, thus indicating the presence of $Hemin/SnO_2$ layers.

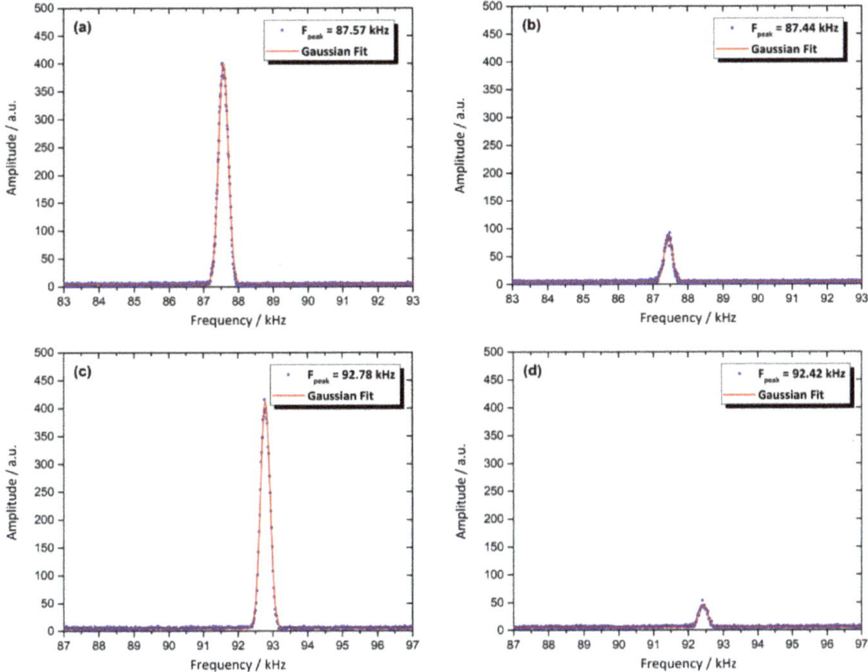

Figure 4. Resonance peaks before and after the immersion of the sensor into the solution for two different samples: (**a**) Bare Metglas ribbon before the immersion; (**b**) Bare Metglas ribbon after the immersion; (**c**) Hemin/SnO_2-Metglas sensor before the immersion; (**d**) Hemin/SnO_2-Metglas sensor after the immersion.

3. Results and Discussion

3.1. X-ray Diffraction

The crystal structure and phase purity of the SnO_2-Metglas film electrode were investigated by the X-ray diffraction technique. Figure 5 shows the main diffraction peaks for Metglas with or without a film of SnO_2 on its surface. The Metglas strip is an amorphous material and therefore its XRD pattern gives a noisy and broad signal with a wide peak from 40° to 50°. The XRD data of the SnO_2 film on Metglas revealed peaks at 26.55°, 33.82°, 37.75° and 51.76°, corresponding to the indices of (101), (110), (200) and (211), are characteristic of the cassiterite type of tetragonal rutile nanocrystals and are consistent with the reported values of the relevant JCPDS card No.: 41-1445 [51]. However, it should be noted that the intensity of the peaks is quite low, which makes us affirm that we have a quite thin film of SnO_2 deposited onto the Metglas strip. Further discussion about the thickness of the deposited film will be given from the SEM images of the SnO_2-Metglas and particularly the cross sectional one which will be used to measure the thickness of the SnO_2 film. Consistent with SEM observation and previous work of ours in the literature [28] the immobilization of hemin did not change the crystal structure of SnO_2.

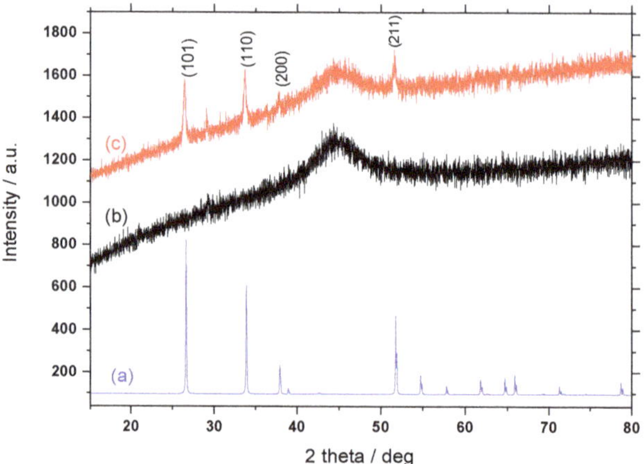

Figure 5. XRD patterns: (**a**) JCPDS card#41-1445 SnO_2 cassiterite; (**b**) Metglas; (**c**) SnO_2/Metglas.

3.2. FE-SEM Imaging of Surface Topography

The surface morphology and thickness of the SnO_2-Metglas film electrode was analyzed by FE-SEM. The top view FE-SEM images presented in Figure 6a,b showed that the SnO_2 film comprises a rigid, porous network of SnO_2 nanoparticles of average size 20–70 nm that are evenly distributed, creating a mesoporous surface. These results confirm a sponge like structure of the film with pore sizes sufficiently large for hemin molecules to be able to diffuse throughout the porous mesostructure. Therefore, it could provide many active sites for catalytic reactions. It was also observed (image not shown) that the immobilization of hemin does not change or destroy the characteristic SnO_2 particles, neither the mesoporous structure of the film. According to Figure 6c the thickness of the SnO_2 films is around ~11 μm as set by the adhesive tape used. Figure 6d displays the EDS for a SnO_2-Metglas film electrode, carried out during the FE-SEM analysis, which conforms to the characteristic peaks of Sn and O. The observed Ni and Fe peaks are attributed to the Metglas substrate. The inset of Figure 6d displays the elemental compositions in estimated weight percentages. The quantification of the EDS spectrum was carried out using the standardless ZAF correction method.

(**a**)   (**b**)

Figure 6. *Cont.*

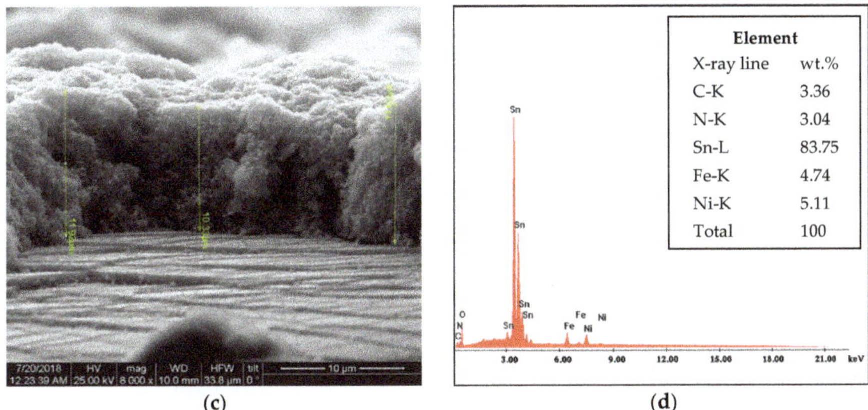

Figure 6. FE-SEM images: (**a**,**b**) Top view; (**c**) cross section of a SnO$_2$-Metglas film electrode; (**d**) EDS elemental microanalysis of a SnO$_2$-Metglas film electrode. Scales are indicated on the photographs.

3.3. Photoluminescence Properties

Figure 7 depicts the photoluminescence spectrum of a SnO$_2$ film deposited on the Metglas substrate at an excitation wavelength of 300 nm. It exhibited four emission bands in UV and visible regions centered around 378, 395, 437 and 470 nm. The 378 band is assigned to the recombination of donor-acceptor pairs [52]. Violet emission i.e., 395 and 437 nm are ascribed to the surface dangling bonds or oxygen vacancies and Sn interstitials [53]. Their exact origin is not yet clear. The origin of blue luminescence at 470 nm is structural defects or impurities formed during the deposition of thin films. It is important to note that the broad nature of the PL emission band clearly indicates that the luminescence should have originated from the multiple sources rather than a single source, because the PL band contains multiple PL peaks.

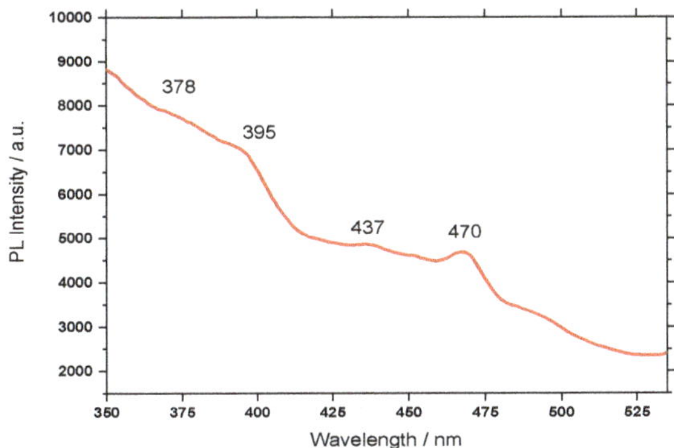

Figure 7. PL spectrum recorded using an excitation wavelength at 300 nm over the SnO$_2$ film deposited on the Metglas 2826MB strip.

3.4. Fourier Transform Infrared Analysis

Figure 8 compares the FTIR spectra of SnO$_2$ and Hemin/SnO$_2$ films recorded in KBr matrices. Blank KBr pellet was taken as the background. Both spectra displayed bands at 3427 and 1638 cm^{-1} due to O–H bonds of adsorbed water and a band located at around 612 cm^{-1}, which is assigned to the Eu mode of SnO$_2$ (anti-symmetric O–Sn–O stretching) [28,54].

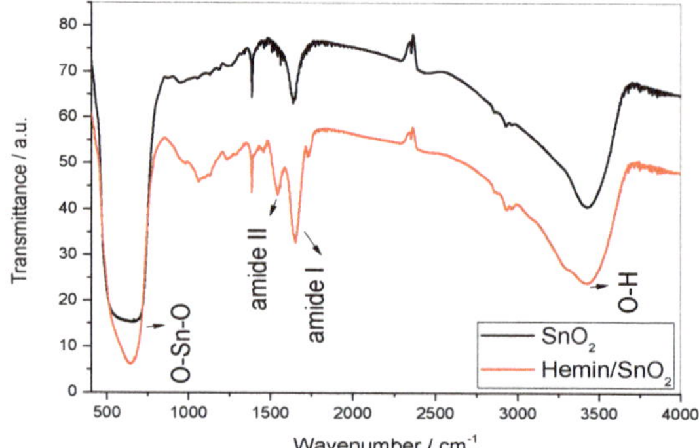

Figure 8. FTIR spectra of SnO_2 (black) and Hemin-SnO_2 (red).

In comparison, the FTIR spectrum of Hemin/SnO_2 displayed the representative peaks of hemin besides the absorption bands of SnO_2. The FTIR spectrum of Hemin/SnO_2 displays new absorption bands at 1000 to 1300 cm^{-1} assigned to C–O stretching vibration in the aromatic ring of the hemin molecule. The small bands at 1410 and 1464 cm^{-1} were due to the contribution of the C–H bending vibration and the band at 2924 cm^{-1} was assigned to the C–H stretching vibration of methyl [55]. The obvious absorption peaks at 1562 and 1651 cm^{-1} must be assigned to amide band II and amide band I, respectively. Therefore, the immobilization of hemin on the SnO_2 mesopores was successfully confirmed by FTIR spectroscopic analysis.

3.5. Electrochemical Behavior of Hemin/SnO_2-Metglas and SnO_2-Metglas Electrodes

Figure 9 shows the CVs of Metglas, SnO_2-Metglas and hemin-modified SnO_2-Metglas in deaerated hemin free 10 mM NaH_2PO_4 (pH = 7) at a scan rate of 0.1 V s^{-1}. In Figure 9a, the CV of the bare Metglas exhibits one broad cathodic peak at −0.4 V and one anodic peak at 0 V. The voltage range is taken from −1 to 0.4 V and in this range the Metglas electrode is effectively working without any breakdown. These peaks could possibly be due to the high content of iron in Metglas which is an amorphous metallic material and thus the iron atoms can occur in both Fe^{2+} and Fe^{3+} states, depending on their local neighborhood in the amorphous atomic framework. These peaks could be a sum of contribution of various oxidation processes of iron to form divalent or trivalent species [56].

The SnO_2-Metglas electrode in Figure 9b shows the characteristic charging/discharging currents assigned to electron injection into sub-band gap/conduction band states of the SnO_2 film as observed previously on different substrates such as fluorine doped tin oxide, FTO glass and indium tin oxide-poly (ethylene terephthalate), ITO-PET [32,50]. The charging of the SnO_2-Metglas electrode starts at +0.1 V, which is very close to the values reported for SnO_2 on other substrates [32,50]. However, the charging starts at a much more positive bias compared with the ZnO film on the same Metglas ribbon we used in a recent study [43], where the charging starts at −0.16 V. In addition, the preparation method of the SnO_2 films compared to the ZnO films used in the past is simpler, faster, of lower cost, using a low temperature route and not a sintering process.

In addition, Figure 9c shows the CV of a Metglas-SnO_2 film after the immobilization of hemin on its surface. This time a couple of strong redox peaks are observed at −0.31 (cathodic) and −0.06 V (anodic). These redox peak currents resulted from the mono-electron transfer process to the immobilized hemin for the conversion from Fe^{3+} to Fe^{2+}. The midpoint potential is estimated form the values of the anodic and cathodic peak potentials, −0.185 V, close to those obtained with hemin on TiO_2 or various carbon electrodes [28,33]. In addition, this value shows a shift to more positive potentials compared with the −0.26 V midpoint potential obtained for hemin on similar SnO_2 films on ITO-PET substrate [32].

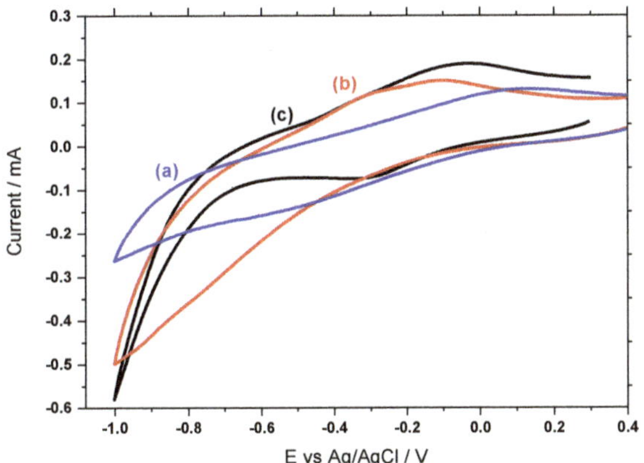

Figure 9. CVs of (**a**) Metglas, (**b**) SnO$_2$-Metglas, (**c**) Hemin/SnO$_2$-Metglas in argon-saturated 10 mM NaH$_2$PO$_4$ (pH = 7) at scan rate: 0.1 V s^{-1}.

It is clear that the mesoporous structure of the SnO$_2$ film not only allows the effective immobilization of hemin on its surface but also promotes the electron transfer process between hemin and electrode at mild applied biases comparable to or even lower than those applied on other electrodes.

Furthermore, the effect of the potential scan rate applied to the Hemin/SnO$_2$-Metglas electrode is investigated and presented in Figure 10a,b. The cathodic and anodic peak currents corresponding to the immobilized hemin vary linearly with the scan rates from 0.1 to 0.01 V s^{-1} (Figure 10b) indicative of a surface-confined electrochemical process. Reduction peaks occur at around -0.31 V and reoxidation at -0.05 V. As the scan rate is steadily increased, the current increases and is significantly bigger if we compare the 0.1 V s^{-1} to the 0.01 V s^{-1}.

The Fe^{3+}/Fe^{2+} redox chemistry of heme is termed quasi-reversible as the peak-to-peak potential separation (ΔE_p) > 60 mV. The ΔE_p is 80 mV at a scan rate of 0.025 V s^{-1} similar to other hemin-modified electrodes reported in the past [28,32,33]. This implies a relatively fast direct electron transfer between the redox active center of hemin and the SnO$_2$-Metglas electrode. The kinetics of the electron transfer for the Hemin/SnO$_2$-Metglas electrode were analyzed using the model of Laviron [57].

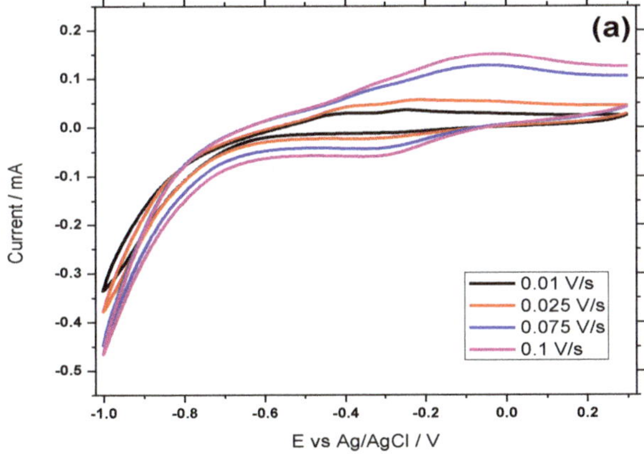

Figure 10. *Cont.*

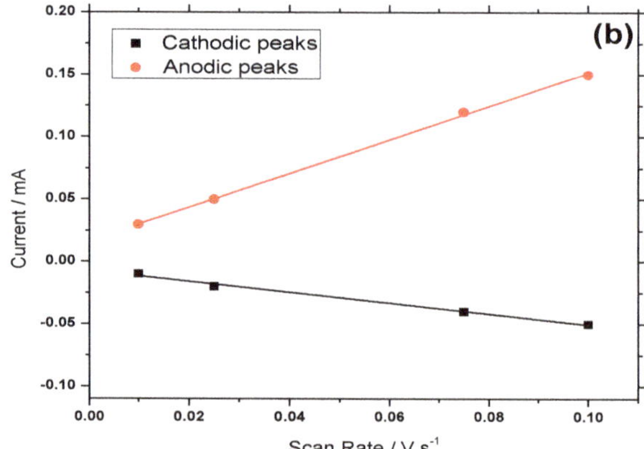

Figure 10. (**a**) CVs of Hemin/SnO$_2$-Metglas films in 10 mM NaH$_2$PO$_4$ buffer (pH = 7) at different scan rates, and (**b**) plot of redox peak currents vs. the scan rates.

A graph of the dependence of peak separation on the logarithm of the scan rate yields a straight line with a slope equal to a charge-transfer coefficient α of 0.88. For a peak separation 0.302 mV and for a scan rate of 0.1 V s^{-1}, a k_s value was estimated to be 0.48 s^{-1}. This value is in the range of k_s for typical surface-controlled quasi-reversible electron transfer. For lower scan rates (less than 0.1 V s^{-1}), the peak separation is decreased, getting closer to the 0 mV expected for an ideal surface controlled reversible electrochemical process. This behavior was achieved without the addition of any promoters or mediators in the electrolyte solution. It is, therefore, obvious that the mesoporous structure of the SnO$_2$ film promotes the rapid electron transfer between hemin and the underlying Metglas surface.

3.6. Electrocatalytic Behavior of H$_2$O$_2$ at the Hemin/SnO$_2$-Metglas Electrode

The suggested simplified mechanism of the electrochemical catalytic reduction of H$_2$O$_2$ on the immobilized hemin can be expressed as the following:

$$\text{Hemin (Fe}^{3+}) + \text{H}^+ + \text{e}^- \rightarrow \text{Heme (Fe}^{2+})$$

$$\text{Heme (Fe}^{2+}) + \text{H}_2\text{O}_2 + 2\text{H}^+ + \text{e}^- \rightarrow \text{Hemin (Fe}^{3+}) + 2\text{H}_2\text{O}$$

Figure 11a displays the electrocatalytic activity of the immobilized hemin toward H$_2$O$_2$ reduction. CV was employed over a range from +0.2 to −1 V. As the increasing concentrations of H$_2$O$_2$ (8–72 μM) were added successively, every 30 s, in the electrochemical cell where our Hemin/SnO$_2$-Metglas electrode was immersed in NaH$_2$PO$_4$ (pH = 7) buffer, the cathodic peak current at −0.49 V increased progressively, indicating the occurrence of the typical electrocatalytic reduction process of H$_2$O$_2$. Simultaneously, the oxidation peak currents decreased accordingly, exhibiting a good electrocatalytic behavior. The reduction potential of H$_2$O$_2$ on our electrodes is around the values reported in the literature [32,43]. Figure 11b shows the proportional, linear (R = 0.987) steady increase of the electrocatalytic cathodic current upon increasing additions of H$_2$O$_2$ in the NaH$_2$PO$_4$ buffer. This plot shows that the CV method is sensitive enough to detect electrochemical changes that occur between the immobilized hemin and H$_2$O$_2$.

Control CVs of hemin free SnO$_2$-Metglas electrodes exhibited negligible dependence upon H$_2$O$_2$ concentration. For comparison, the peak current at the cathodic peak at −0.49 V of the Hemin/SnO$_2$-Metglass electrode is plotted in Figure 12 as solid squares, together with the corresponding signal (solid triangles) which is obtained when H$_2$O$_2$ is added in the electrolyte solution. The control experiment produced a flat response with a small error of 0.01 mA, and it is obvious that the changes brought up with the reduction of H_2O_2 by the immobilized hemin produced a big enough sensing signal of about 0.15 mA in variation, much larger than the above error, confirming the good sensitivity of the CV method.

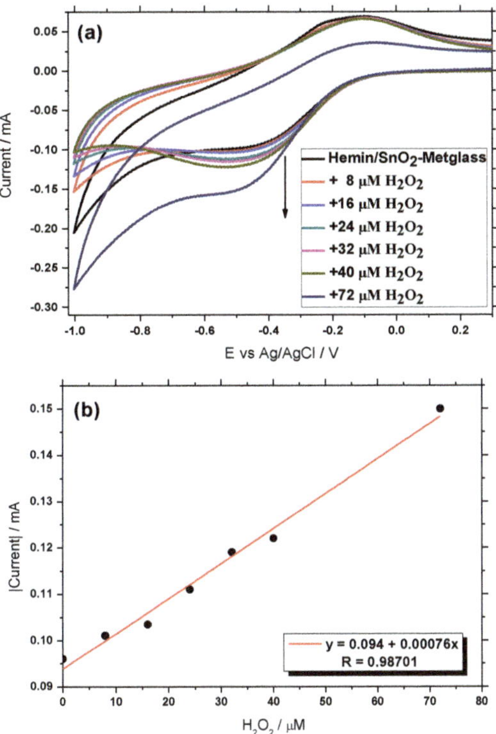

Figure 11. (**a**) CVs of Hemin/SnO$_2$-Metglas in the absence and presence of increasing concentrations (8–72 µM) of H$_2$O$_2$ in 10 mM NaH$_2$PO$_4$ (pH = 7), scan rate: 0.1 V s^{-1}, and (**b**) plot of the cathodic peak current vs. H$_2$O$_2$ concentration obtained from the CV data shown in (**a**).

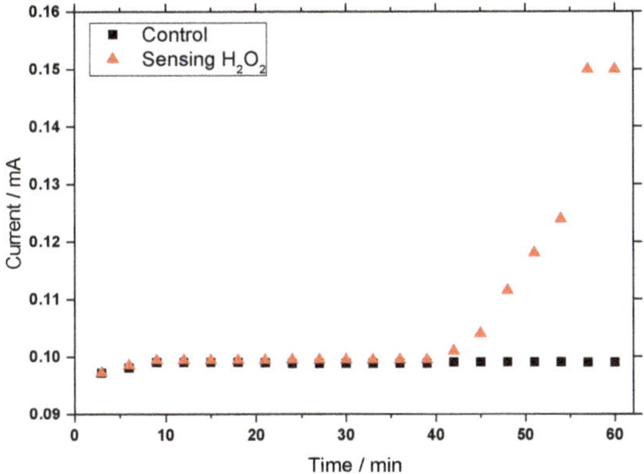

Figure 12. Comparison of the control cathodic peak currents (squares) obtained from the CVs of a Hemin/SnO$_2$-Metglas electrode in 10 mM NaH$_2$PO$_4$ solution and the corresponding sensing current (triangles) when H$_2$O$_2$ is added in the solution.

3.7. Amperometric Sensing of H_2O_2

Based on the electrocatalytic results above, a biosensor for the quantitative determination of H_2O_2 is proposed. To improve the sensitivity, the performance of the Hemin/SnO$_2$-Metglas film electrode towards the determination of H_2O_2 was evaluated by amperometry. Figure 13a shows a typical current-time curve of a Hemin/SnO$_2$-Metglas film electrode for the successive additions of H_2O_2 in a stirred cell with NaH$_2$PO$_4$ buffer and an applied potential at -0.4 V. The sensor gives continuous real-time current responses to changing H_2O_2 concentrations. The reduction peak potential for H_2O_2 displayed in Figure 11a (CV measurements) is around -0.5 V. However, the applied potential for amperometric sensing of H_2O_2 should be more positive in order to decrease the background current and minimize the response/effect of common interferents. From our study of the influence of the applied potential on the amperometric response to H_2O_2 we concluded that a -0.3 V produced a very low response and the signal was enhanced at -0.4 V. When the applied potential was further negatively shifted up to -0.6 V, the background current increased but the current response toward H_2O_2 decreased. Therefore, -0.4 V was selected as the applied potential for our chronoamperometric sensor.

Figure 13a shows that after the addition of H_2O_2 significant increases in the cathodic current are observed. The response reaches the steady-state value within 10–15 s which is a relatively fast response. Figure 13b corresponds to the calibration plot of the prepared biosensor. The currents had a linear dependence on the concentration of H_2O_2 in the range of 2–90 μM with a correlation coefficient 0.995 and the detection limit was examined to be 1.6×10^{-7} M.

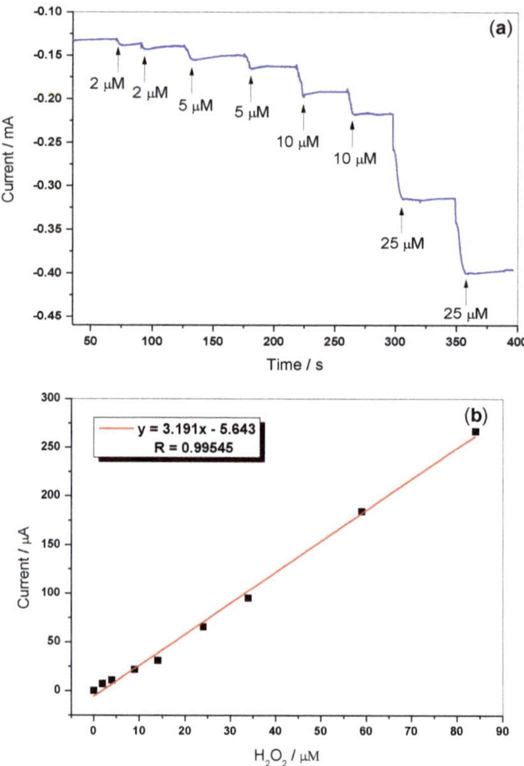

Figure 13. (**a**) The typical amperometric responses of a Hemin/SnO$_2$-Metglas film on successive additions of different amounts of H_2O_2 into stirring NaH$_2$PO$_4$ (10 mM, pH = 7) saturated with Argon at the applied potential of -0.4 V. (**b**) The calibration curve of the current vs. the H_2O_2 concentration.

The analytical performance of our Hemin/SnO_2-Metglass electrode towards H_2O_2 is compared with other hemin-modified electrodes which were used in the past for the development of amperometric and/or voltammetric sensors and the results are summarized in Table 2. It is obvious that the sensitivity (linear range) of our sensor compares well with the other electrodes that have been used for the development of H_2O_2 sensors. It is also notable that the detection limit (LOD) of our electrode is superior in some cases as compared to the previously reported electrode materials. These very good results obtained for our sensor are due to the high active area of the SnO_2 film on the Metglas substrate, the high hemin loading achieved in a stable and functional way, and the fact that fast and direct electron transfer is achieved between the immobilized hemin and the SnO_2-Metglas electrode without the use of any mediators or promoters. However, the aim was not to produce the most sensitive electrochemical sensor for H_2O_2, but rather one which will compare well with other electrodes and would be more sensitive that the Hb/ZnO-Metglas sensor we developed in the past. A much lower LOD for H_2O_2 was obtained for this sensor compared with the Hb/ZnO-Metglas electrodes.

Table 2. Comparison of the analytical performance with different electrode materials for the electrocatalytic determination of H_2O_2.

Electrode	Method	Linear Range (M)	LOD (M)	Ref.
Hemin/SnO_2-ITO/PET	CV	1.5×10^{-6}–90×10^{-6}	1.5×10^{-6}	[32]
Hb/ZnO-Metglas	CV & MR	25×10^{-6}–350×10^{-6}	25×10^{-6}–50×10^{-6}	[43]
Hemin-graphene nano-sheets (H-GNs)/gold nano-particles (AuNPs)	CV & Amperometry	0.3×10^{-6}–1.8×10^{-3}	0.11×10^{-6}	[58]
Hemin-TiO_2 modified electrode	CV & Amperometry	3.0×10^{-7}–4.7×10^{-4}	7.2×10^{-8}	[28]
Hemin-GCE	CV	0–170×10^{-6}	31.6×10^{-6}	[59]
ITO/NiO/Hemin	CV	0.5×10^{-6}–500×10^{-6}	10^{-7}	[26]
Hemin/SnO_2-Metglas	CV & MR	2×10^{-6}–90×10^{-6}	1.6×10^{-7}	This work

Furthermore, the selectivity of this sensor for interferents that may be present in real samples has been tested by our group in a recent study [32]. Our sensor is free of interferences like uric acid (2%) and exhibits a slight interference (7%–8%) of response currents to 100 µM ascorbic acid relative to 100 µM H_2O_2. The reproducibility of our electrodes was also tested. Their preparation method for the SnO_2 films is very simple, involves very few steps and a low-temperature route; therefore, they can be easily scaled-up on the metallic ribbon substrate. Six separate films prepared under the same conditions produced very similar results for the electrochemical reduction of hemin and its electrocatalytic performance towards H_2O_2 with a relative standard deviation of around 4%–5%. After preparation the electrodes could be stored for up to two weeks at 5 °C and could be used repeatedly maintaining at least 90% of their electrocatalytic activity.

3.8. Magnetic Resonance Behavior of the Sensor

The MR measurements, as described in the section on the experimental procedure, were performed on four Metglas ribbons, of which one was bare Metglas, one was SnO_2-Metglas and the last two were Hemin/SnO_2-metglas ribbons. Figure 14 shows the resonance behavior of the ribbons versus time, with Figure 14a–c, corresponding to bare Metglas, SnO_2-Metglas and Hemin/SnO_2-Metglas in the presence of 72 µM H_2O_2 in the solution, respectively. Figure 14d, on the other hand, shows a Hemin/SnO_2-Metglas without H_2O_2 present in the solution. For each ribbon, a total number of 30 measurements were performed, each one every 3 min. A change of 0.09 ± 0.01 kHz was observed for the Hemin/SnO_2-Metglas ribbon with H_2O_2 (Figure 14c), indicating the mass increase on the sensor. For the particular Metglas ribbon used (length = 2.5 cm), calibration with known small mass loads gives a calibration factor of -1.63 kHz·mg^{-1}. Using this factor, the maximum H_2O_2 concentration of 72 µM and the resonance frequency change, a corresponding mass increase of 767 ng/µM was calculated on the sensor. On the other hand, for the Hemin/SnO_2-Metglas ribbon without H_2O_2 in the solution (Figure 14d), almost no changes occurred on the resonance frequencies, showing thus the effect of H_2O_2 on the sensor. The control diagrams of Figure 14a,b for bare Metglas and SnO_2-Metglas showed no significant resonance changes in the presence of H_2O_2, as the interaction element, hemin, is not immobilized on their surfaces. However, Figure 14b shows a slight resonance change. This happens due to the porous structure of the SnO_2 film, which can encapsulate some H_2O_2 molecules present in the solution and therefore increase slightly the mass on the sensor.

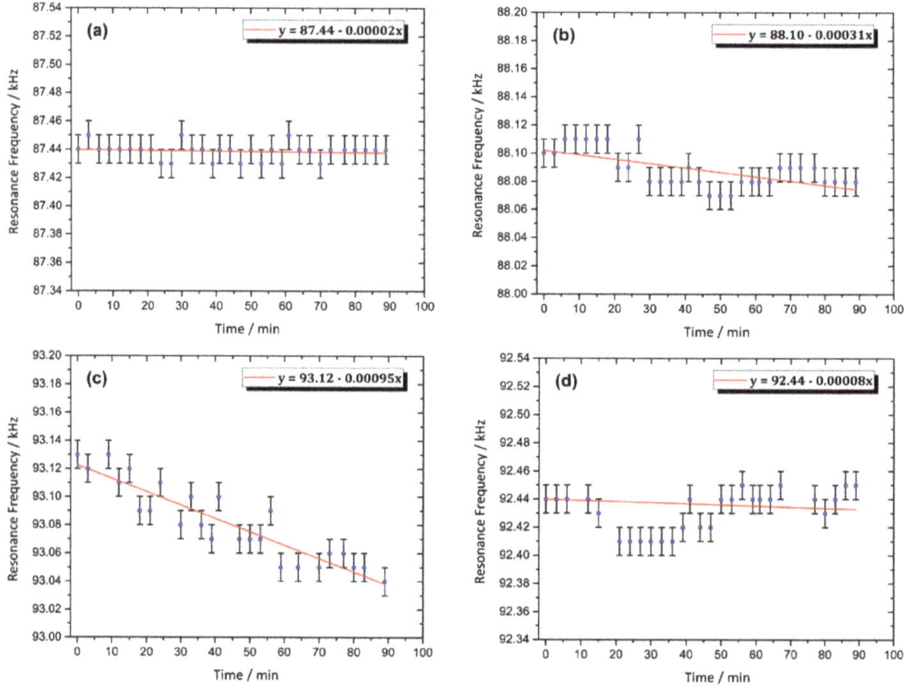

Figure 14. Magnetic resonance measurements of (**a**) Bare Metglas ribbon with H_2O_2, (**b**) SnO_2-Metglas sensor with H_2O_2, (**c**) Hemin/SnO_2-Metglas sensor with H_2O_2, (**d**) Hemin/SnO_2-Metglas sensor without H_2O_2.

4. Conclusions

In this work, we reported a simple, low-temperature method for preparing Hemin/SnO_2 films on a magnetic ribbon substrate for the detection of H_2O_2. Two detection methods, CV and MR were used successfully for the simultaneous sensing of H_2O_2. High hemin loading in a stable and functional way, using a simple coating method, was achieved. Direct and fast electron transfer between the immobilized hemin and the SnO_2-Metglas electrode was observed without the use of any mediators or promoters. A much lower LOD for H_2O_2 was obtained compared with the Hemin/ZnO-Metglas electrodes we had used in the past. Experimental MR results showed a sensible change of the resonance frequency of the Hemin/SnO_2-Metglas sensor in the presence of H_2O_2, confirming thus the mass change on the sensor. Specifically, a calculated mass increase of about 767 ng/µM was achieved after the addition of H_2O_2, much higher than the 152 ng/µM obtained for the Hemin/ZnO-Metglas electrodes in the past. It should also be noted that no interference occurred when both techniques were applied on the same Hemin/SnO_2-Metglas electrode. This approach should be extendable to many other analytes of interest in clinical control, environmental, food and industrial analysis that have been detected so far using mostly sensitive electrochemical techniques. MR could be applied for the sensing of these analytes simultaneously, confirming the sensitivity of the electrochemical methods and developing novel magnetoelastic sensors.

Author Contributions: E.T. and G.S. conceived and designed the experiments; P.N., G.S., A.G.-L., E.S. and I.M.B. performed the experiments; E.T., G.S. and P.N. analyzed the data; E.T. and G.S. contributed reagents/materials/analysis tools; E.T., G.S. and P.N. wrote the paper.

Funding: This research received no external funding.

Conflicts of Interest: The authors declare no conflict of interest.

References

1. Woo, Y.A.; Lim, H.R.; Kim, H.J.; Chung, H. Determination of hydrogen peroxide concentration in antiseptic solutions using portable near-infrared system. *J. Pharm. Biomed. Anal.* **2003**, *33*, 1049–1057. [CrossRef]

2. Wang, T.Y.; Zhu, H.C.; Zhuo, J.Q.; Zhu, Z.W.; Papakonstantinou, P.; Lubarsky, G.; Lin, J.; Li, M.X. Biosensor Based on Ultrasmall MoS_2 Nanoparticles for Electrochemical Detection of H_2O_2 Released by Cells at the Nanomolar Level. *Anal. Chem.* **2013**, *85*, 10289–10295. [CrossRef] [PubMed]

3. Bai, J.; Jiang, X. A Facile One-Pot Synthesis of Copper Sulfide-Decorated Reduced Graphene Oxide Composites for Enhanced Detecting of H_2O_2 in Biological Environments. *Anal. Chem.* **2013**, *85*, 8095–8101. [CrossRef] [PubMed]

4. Silva, R.A.B.; Montes, R.H.O.; Richter, E.M.; Munoz, R.A.A. Rapid and selective determination of hydrogen peroxide residues in milk by batch injection analysis with amperometric detection. *Food Chem.* **2012**, *133*, 200–204. [CrossRef]

5. Seders, L.A.; Shea, C.A.; Lemmon, M.D.; Maurice, P.A.; Talley, J.W. LakeNet: An Integrated Sensor Network for Environmental Sensing in Lakes. *Environ. Eng. Sci.* **2007**, *24*, 183–191. [CrossRef]

6. Takahashi, A.; Hashimoto, K.; Kumazawa, S.; Nakayama, T. Determination of hydrogen peroxide by high-performance liquid chromatography with a cation-exchange resin gel column and electrochemical detector. *Anal. Sci.* **1999**, *15*, 481–483. [CrossRef]

7. Matsubara, C.; Kawamoto, N.; Takamura, K. Oxo [5, 10, 15, 20-tetra(4-pyridyl) porphyrinato] titanium (IV): An ultra-high sensitivity spectrophotometric reagent for hydrogen peroxide. *Analyst* **1992**, *117*, 1781–1784. [CrossRef]

8. Zscharnack, K.; Kreisig, T.; Prasse, A.A.; Zuchner, T. A luminescence-based probe for sensitive detection of hydrogen peroxide in seconds. *Anal. Chim. Acta* **2014**, *834*, 51–57. [CrossRef] [PubMed]

9. Xu, C.; Ren, J.; Feng, L.; Qu, X. H_2O_2 triggered sol–gel transition used for visual detection of glucose. *Chem. Commun.* **2012**, *48*, 3739–3741. [CrossRef] [PubMed]

10. Sun, J.; Li, C.; Qi, Y.; Guo, S.; Liang, X. Optimizing Colorimetric Assay Based on V_2O_5 Nanozymes for Sensitive Detection of H_2O_2 and Glucose. *Sensors* **2016**, *16*, 584. [CrossRef] [PubMed]

11. Kim, J.H.; Patra, C.R.; Arkalgud, J.R.; Boghossian, A.A.; Zhang, J.H.; Han, J.; Reuel, N.F.; Ahn, J.H.; Mukhopadhyay, D.; Strano, M.S. Single-Molecule Detection of H_2O_2 Mediating Angiogenic Redox Signaling on Fluorescent Single-Walled Carbon Nanotube Array. *ACS Nano* **2011**, *5*, 7848–7857. [CrossRef] [PubMed]

12. Shamsipur, M.; Pashabadi, A.; Molaabasi, F. A novel electrochemical hydrogen peroxide biosensor based on hemoglobin capped gold nanoclusters-chitosan composite. *RSC Adv.* **2015**, *5*, 61725–61734. [CrossRef]

13. Xu, Y.; Hu, C.; Hu, S. A hydrogen peroxide biosensor based on direct electrochemistry of hemoglobin in Hb-Ag sol films. *Sens. Actuators B* **2008**, *130*, 816–822. [CrossRef]

14. Duan, G.; Li, Y.; Wen, Y.; Ma, X.; Wang, Y.; Ji, J.; Wu, P.; Zhang, Z.; Yang, H. Direct electrochemistry and electrocatalysis of Hemoglobin/ZnO-Chitosan/nano-Au modified glassy carbon electrode. *Electroanalysis* **2008**, *20*, 2454–2459. [CrossRef]

15. Zhu, C.; Yang, G.; Li, H.; Du, D.; Lin, Y. Electrochemical sensors and biosensors based on nanomaterials and nanostructures. *Anal. Chem.* **2015**, *85*, 230–249. [CrossRef] [PubMed]

16. Liu, X.; Ma, T.; Pinna, N.; Zhang, J. Two-dimensional nanostructured materials for gas sensing. *Adv. Funct. Mater.* **2017**, 1702168. [CrossRef]

17. Zhang, J.; Liu, X.; Neri, G.; Pinna, N. Nanostructured material for room-temperature gas sensors. *Adv. Mater.* **2016**, *28*, 795–831. [CrossRef] [PubMed]

18. Joshi, N.; Hayasaka, T.; Liu, Y.; Liu, H.; Oliveira Jr, O.N.; Lin, L. A review on chemiresistive room temperature gas sensors based on metal oxide nanostructures, graphene and 2D transition metal dichalcogenides. *Microchim. Acta* **2018**, *185*, 213–228. [CrossRef] [PubMed]

19. Zhang, Y.Y.; Bai, X.Y.; Wang, X.M.; Shiu, K.K.; Zhu, Y.L.; Jiang, H. Highly Sensitive Graphene-Pt Nanocomposites Amperometric Biosensor and Its Application in Living Cell H_2O_2 Detection. *Anal. Chem.* **2014**, *86*, 9459–9465. [CrossRef] [PubMed]

20. Xue, S.; Jing, P.; Xu, W. Hemin on graphene nanosheets functionalized with flower-like MnO_2 and hollow AuPd for the electrochemical sensing lead ion based on the specific DNAzyme. *Biosens. Bioelectron.* **2016**, *86*, 958–965. [CrossRef] [PubMed]

21. Wang, L.; Yang, H.; He, J.; Zhang, Y.; Yu, J.; Song, Y. Cu-Hemin Metal Organic Frameworks Chitosan Reduced Graphene Oxide Nanocomposites with Peroxidase-Like Bioactivity for Electrochemical Sensing. *Electrochim. Acta* **2016**, *213*, 691–697. [CrossRef]

22. Yang, X.; Xiao, F.B.; Lin, H.W.; Wu, F.; Chen, D.Z.; Wu, Z.Y. A novel H_2O_2 biosensor based on Fe_3O_4–Au magnetic nanoparticles coated horseradish peroxidase and graphene sheets–Nafion film modified screen-printed carbon electrode. *Electrochim. Acta* **2013**, *109*, 750–755.

23. Lv, X.C.; Weng, J. Ternary Composite of Hemin, Gold Nanoparticles and Graphene for Highly Efficient Decomposition of Hydrogen Peroxide. *Sci. Rep.* **2013**, *3*, 3285. [CrossRef] [PubMed]

24. Wang, J.; Chen, X.; Liao, K.; Wang, G.; Han, M. Pd nanoparticle-modified electrodes for nonenzymatic hydrogen peroxide detection. *Nanoscale Res. Lett.* **2015**, *10*, 311. [CrossRef] [PubMed]

25. Chang, G.H.; Luo, Y.L.; Lu, W.B.; Qin, X.Y.; Sun, X.P. Carbon nanoparticles-induced formation of polyaniline nanofibers and their subsequent decoration with Ag nanoparticles for nonenzymatic H_2O_2 detection. *Russ. J. Electrochem.* **2014**, *50*, 95–99. [CrossRef]

26. Gu, T.T.; Wu, X.M.; Dong, Y.M.; Wang, G.L. Novel photoelectrochemical hydrogen peroxide sensor based on hemin sensitized nanoporous NiO based photocathode. *J. Electoanal. Chem.* **2015**, *759*, 27–31. [CrossRef]

27. Huang, J.F.; Zhu, Y.H.; Zhong, H.; Yang, X.; Li, C.Z. Dispersed CuO nanoparticles on a silicon nanowire for improved performance of nanoenzymatic H_2O_2 detection. *ACS Appl. Mater. Inter.* **2014**, *6*, 7055–7062. [CrossRef] [PubMed]

28. Zhu, Y.; Yan, K.; Xu, Z.; Zhang, J. Hemin Modified TiO_2 Nanoparticles with Enhanced Photoelectrocatalytic Activity for Electrochemical and Photoelectrochemical Sensing. *J. Electrochem. Soc.* **2016**, *163*, B526–B532. [CrossRef]

29. Huan, Y.F.; Fei, Q.; Shan, H.Y.; Wang, B.J.; Xu, H.; Feng, G.D. A novel water-soluble sulfonated porphyrin fluorescence sensor for sensitive assays of H_2O_2 and glucose. *Analyst* **2015**, *140*, 1655–1661. [CrossRef] [PubMed]

30. Wu, H.; Wei, T.; Li, X.; Yang, J.; Zhang, J.; Fan, S.; Zhang, H. Synergistic-Effect-Controlled Tetraoctylammonium Bromide/Multi-Walled Carbon Nanotube@Hemin Hybrid Material for Construction of Electrochemical Sensor. *J. Electrochem. Soc.* **2017**, *164*, B147–B151. [CrossRef]

31. Cao, Y.; Si, W.; Hao, Q.; Li, Z.; Lei, W.; Xia, X.; Li, J.; Wang, F.; Liu, Y. One–pot fabrication of Hemin-N-C composite with enhanced electrocatalysis and application of H_2O_2 sensing. *Electrochim. Acta* **2018**, *261*, 206–213. [CrossRef]

32. Panagiotopoulos, A.; Gkouma, A.; Vassi, A.; Johnson, C.J.; Cass, A.E.G.; Topoglidis, E. Hemin modified SnO_2 films on ITO-PET with enhanced activity for electrochemical sensing. *Electroanalysis* **2018**. in Press. [CrossRef]

33. Wang, Y.; Hosono, T.; Hasebe, Y. Hemin-adsorbed carbon felt for sensitive and rapid flow-amperometric detection of dissolved oxygen. *Microchim. Acta* **2013**, *180*, 1295–1302. [CrossRef]

34. Santos, R.M.; Rodrigues, M.S.; Laranjinha, J.; Barbosa, R.M. Biomimetic sensor based on hemin/carbon nanotubes/chitosan modified microelectrode for nitric oxide measurement in the brain. *Biosens. Bioelectron.* **2013**, *44*, 152–159. [CrossRef] [PubMed]

35. Pavithra, L.; Devasena, T.; Pandian, K.; Gopinath, S.C.B. Amperometric determination of nitrite using natural fibers as template for titanium dioxide nanotubes with immobilized hemin as electron transfer mediator. *Microchim. Acta* **2018**, *185*, 194.

36. Garcia de la Rosa, A.; Castro-Quezada, E.; Gutierrez-Granados, S.; Bedioui, F.; Alatorre-Ordaz, A. Stable hemin embedded in Nafion films for the catalytic reduction of trichloroacetic acid under hydrodynamic conditions. *Electrochem. Commun.* **2005**, *7*, 853–856.

37. Obare, S.O.; Ito, T.; Balfour, M.H.; Meyer, G.J. Ferrous hemin oxidation by organic halides at nanocrystalline TiO_2 interfaces. *Nano Lett.* **2003**, *3*, 1151–1153. [CrossRef]

38. Zhang, T.; Wang, L.; Gao, C.; Zhao, C.; Wang, Y.; Wang, J. Hemin immobilized into metal-organic frameworks as electrochemical biosensor for 2,4,6-trichlorophenol. *Nanotechnology* **2018**, *29*, 074003. [CrossRef] [PubMed]

39. Reys, J.R.M.; Lima, P.R.; Cioletti, A.G.; Ribeiro, A.S.; Abreu, F.C.D.; Goulart, M.O.F.; Kubota, L.T. An amperometric sensor based on hemin adsorbed on silica gel modified with titanium oxide for electrocatalytic reduction and quantification of artemisinin. *Talanta* **2008**, *77*, 909–914. [CrossRef]
40. Guo, Y.; Li, J.; Dong, S.J. Hemin functionalized graphene nanosheets-based dual biosensor platforms for hydrogen peroxide and glucose. *Sens. Actuators B* **2011**, *160*, 295–300. [CrossRef]
41. Tao, Y.; Ju, E.G.; Ren, J.S.; Qu, X.G. Polypyrrole nanoparticles as promising enzyme mimics for sensitive hydrogen peroxide detection. *Chem. Commun.* **2014**, *50*, 3030–3032. [CrossRef] [PubMed]
42. Wang, J.J.; Han, D.X.; Wang, X.H.; Qi, B.; Zhao, M.S. Polyoxometalates as peroxidase mimetics and their applications in H_2O_2 and glucose detection. *Biosens. Bioelectron.* **2012**, *36*, 18–21. [CrossRef] [PubMed]
43. Sagasti, A.; Bouropoulos, N.; Kouzoudis, D.; Panagiotopoulos, A.; Topoglidis, E.; Gutierrez, J. Nanostructured ZnO in a Metglas/ZnO/Hemoglobin modified electrode to detect the oxidation of Hemoglobin simultaneously by cyclic voltammetry and magnetoelastic resonance. *Materials* **2017**, *10*, 849. [CrossRef] [PubMed]
44. Topoglidis, E.; Cass, A.E.G.; O'Regan, B.; Durrant, J.R. Immobilization and bioelectrochemistry of proteins on nanoporous TiO_2 and ZnO films. *J. Electroanal. Chem.* **2001**, *517*, 20–27. [CrossRef]
45. Willit, J.L.; Bowden, E.F. Adsorption and redox thermodynamics of strongly adsorbed cytochrome c on tin oxide electrodes. *J. Phys. Chem.* **1990**, *94*, 8241–8246. [CrossRef]
46. Zhang, K.; Zhang, L.; Chai, Y. Mass load distribution dependence of mass sensitivity of magnetoelastic sensors under different resonance modes. *Sensors* **2015**, *15*, 20267–20278. [CrossRef] [PubMed]
47. Li, S.; Cheng, Z.Y. Nonuniform mass detection using magnetostrictive biosensors operating under multiple harmonic resonance modes. *J. Appl. Phys.* **2010**, *107*, 114514. [CrossRef]
48. Hernando, A.; Vazquez, M.; Barandiaran, J. Metallic glasses and sensing applications. *J. Phys. E Sci. Instrum.* **1988**, *21*, 1129. [CrossRef]
49. Modzelewski, C.; Savage, H.; Kabacoff, L.; Clark, A. Magnetomechanical coupling and permeability in transversely annealed Metglas 2605 alloys. *IEEE Trans. Magn.* **1981**, *17*, 2837–2839. [CrossRef]
50. Topoglidis, E.; Astuti, Y.; Duriaux, F.; Gratzel, M.; Durrant, J.R. Direct Electrochemistry and Nitric Oxide Interaction of Heme Proteins Adsorbed on Nanocrystalline Tin Oxide Electrodes. *Langmuir* **2003**, *19*, 6894–6900. [CrossRef]
51. *Joint Committee on Powder Diffraction Standards (JCPDS) Card No. 41-1445 for SnO2*; International Center for Diffraction Data: Newtown Township, PA, USA.
52. Sivashankaran, N.S.L.; Vijayam, S.N.A.; PuthenKadathil, V.T.; Kunjkunju, J. Magnetic properties of Mn-doped SnO_2 thin films prepared by the Sol–Gel dip coating method for dilute magnetic semiconductors. *J. Am. Ceram. Soc.* **2014**, *97*, 3184–3191. [CrossRef]
53. Shikha, B.; Pandya, D.K.; Kashyap, S.C. Photoluminescence and Defects in Ultrathin SnO_2 Films. *AIP Conf. Proc.* **2013**, *1512*, 1042–1043. [CrossRef]
54. Seema, H.; Kemp, K.C.; Chandra, V.; Kin, K.S. Graphene–SnO_2 composites for highly efficient photocatalytic degradation of methylene blue under sunlight. *Nanotechnology* **2012**, *23*, 355705. [CrossRef] [PubMed]
55. Jahn, M.R.; Shukoor, I.; Tremel, W.; Wolfrum, U.; Kolb, U.; Naworth, T.; Langguth, P. Hemin-coupled iron (III)-hydroxide nanoparticles show increased uptake in Caco-2 cells. *J. Pharm. Pharmacol.* **2011**, *63*, 1522–1530. [CrossRef] [PubMed]
56. Altube, A.; Pierna, A.R. Thermal and electrochemical properties of cobalt containing Finemet type alloys. *Electrochem. Acta* **2004**, *49*, 303–311. [CrossRef]
57. Laviron, E.J. General expression of the linear potential sweep voltammogram in the case of diffusionless electrochemical systems. *Electroanal. Chem.* **1979**, *101*, 19–28. [CrossRef]
58. Song, H.; Ni, Y.; Kokot, S. A novel electrochemical biosensor based on the hemin-graphene nano-sheets and gold nano-particles hybrid film for the analysis of hydrogen peroxide. *Anal. Chim. Acta* **2013**, *788*, 24–31. [CrossRef] [PubMed]
59. Deac, A.R.; Morar, C.; Turdean, G.L.; Darabentu, M.; Gal, E.; Bende, A.; Muresan, L.M. Glassy carbon electrode modified with hemin and new melamine compounds for H_2O_2 amperometric detection. *J. Solid State Electrochem.* **2016**, *20*, 3071–3081. [CrossRef]

Article

Ethylene Glycol Functionalized Gadolinium Oxide Nanoparticles as a Potential Electrochemical Sensing Platform for Hydrazine and p-Nitrophenol

Savita Chaudhary [1,*], **Sandeep Kumar** [1], **Sushil Kumar** [1], **Ganga Ram Chaudhary** [1], **S.K. Mehta** [1] and **Ahmad Umar** [2,3,*]

[1] Department of Chemistry and Centre of Advanced Studies in Chemistry, Panjab University, Chandigarh 160014, India; sandeepkumar@gmail.com (S.K.); sushilkumarkai@gmail.com (S.K.); grc22@pu.ac.in (G.R.C.); sav66hooda@gmail.com (S.K.M.)

[2] Department of Chemistry, College of Science and Arts, Najran University, Najran 11001, Saudi Arabia

[3] Promising Centre for Sensors and (Electronic Devices PCSED), Najran University, Najran 11001, Saudi Arabia

* Correspondence: schaudhary@pu.ac.in (S.C.); umahmad@nu.edu.sa (A.U.);
   Tel.: +911-722534437 (S.C.); +966-534574597 (A.U.)

Received: 20 July 2019; Accepted: 26 September 2019; Published: 1 October 2019

Abstract: The current work reports the successful synthesis of ethylene glycol functionalized gadolinium oxide nanoparticles (Gd_2O_3 Nps) as a proficient electrocatalytic material for the detection of hydrazine and p-nitrophenol. A facile hydrothermal approach was used for the controlled growth of Gd_2O_3 Nps in the presence of ethylene glycol (EG) as a structure-controlling and hydrophilic coating source. The prepared material was characterized by several techniques in order to examine the structural, morphological, optical, photoluminescence, and sensing properties. The thermal stability, resistance toward corrosion, and decreased tendency toward photobleaching made Gd_2O_3 nanoparticles a good candidate for the electrochemical sensing of p-nitrophenol and hydrazine by using cyclic voltammetric (CV) and amperometric methods at a neutral pH range. The modified electrode possesses a linear range of 1 to 10 µM with a low detection limit of 1.527 and 0.704 µM for p-nitrophenol and hydrazine, respectively. The sensitivity, selectivity, repeatability, recyclability, linear range, detection limit, and applicability in real water samples made Gd_2O_3 Nps a favorable nanomaterial for the rapid and effectual scrutiny of harmful environmental pollutants.

Keywords: gadolinium oxide; hydrazine; p-nitrophenol; electrochemical sensing; amperometric; selective sensor

---

1. Introduction

Aromatic nitro as well as hydrazine are some of the few compounds that are frequently used in the preparation of insecticides, pesticides, pharmaceuticals, and in chemical industries [1–3]. The highly stable nature and lower degradation efficiency of these compounds imparted serious health hazards to human health [4,5]. The utilities of these chemicals in the preparation of explosives are well established in the literature [6]. For instance, according to the U.S Homeland Security Information Bulletin, hydrazine was used in a terrorist attack in 2003 [7]. Hence, from the perspective of safety and security, the development of simple, handy, and competent methodology to monitor these contaminants is the crucial need of the society [8]. To date, there have been a number of analytical instrumental techniques—such as colorimetric, X-ray, fluorescence emission spectroscopy, inductively coupled plasma spectroscopy, atomic absorption spectroscopy, mass spectroscopy, and chromatography—that have been employed for the detection of p-nitrophenol as well as hydrazine [9–12].

All these methods are quite efficient for the detection of these pollutants, but possess delicate functioning, high processing charges, and skilled professionals for data analysis [12]. All these hitches have restricted the use of these sophisticated techniques in routine applications. Hence, from the viewpoint of human health and environmental security, there is a critical requirement for developing alternative techniques with improved selectivity and sensitivity toward these contaminants [13–15]. Therefore, in this work, we have coupled the sensitivity of electrochemical technique with nanoparticles for developing effective sensors for these harmful pollutants. The developed electrochemical sensor has offered significant benefits such as low processing cost and quick response time. The presence of nanoparticles has further augmented the mass transport during analysis as well as reduced the effect of opposition produced by solution during the measurements. The signal-to-noise ratio is further enhanced in the presence of nanoparticles as compared to conventional macroelectrodes used during the analysis. The higher available surface area of nanoparticles has made them an efficient adsorbent for analytes and provided better-quality responses for contaminants.

In the past, varieties of nanoparticles have been used for preparing electrochemical sensors for hydrazine and aromatic compounds [15–20]. For instance, Mishra et al. [15,16] developed the flexible epidermal tattoo, textile and glove-based electrochemical sensor for the detection of organophosphate molecules. The developed sensor was found to be effective in defense and food security applications. Wei et al. [17] have used the nanohybrids of carbon nanotubes (CNTs) with pyrene-cyclodextrin for the electrochemical sensing of p-nitrophenol with sensitivity of around 18.7 $\mu A/\mu M$. Karthik et al. [18] have used the applications of gold nanoparticles derived from biogenic sources for sensing hydrazine from aqueous media. The developed sensor has shown a linear range 5 nM to 272 μM with a detection limit of around 0.05 μM. Zhang and co-workers have used the application of modified graphene with Pt–Pd nanocubes for detecting aromatic nitro compounds with a detection range of 0.01 to 3 ppm and sensing limit of around 0.8 ppb [19]. Chen et al. [20] have developed the indium tin oxide electrodes, which were further functionalized with β-cyclodextrin and Ag nanoparticles for analyzing the trace amount of nitroaromatic compounds via using electrochemical sensing analysis. Chaudhary et al. [21] have utilized the fluorescence sensing abilities of Gd_2O_3 Nps nanoparticles in the selective detection of 4-nitrophenol in aqueous media. However, the use of Gd_2O_3 Nps for the modification of electrodes was less frequent in the literature for the estimation of harmful hydrazine and aromatic compounds.

The current work has utilized the electron transport abilities, high electrical conductivity, and thermal stabilities of Gd_2O_3 Nps for making effective material in electrochemical sensing for harmful pollutants [21–23]. To date, a diverse range of methodologies has been investigated for the preparation of Gd_2O_3 Nps [24–28]. The available methods have generally required very high temperature reaction conditions and multistep processing for the synthesis of Gd_2O_3 Nps. Therefore, it is valuable to systematize a synthetic approach for preparing Gd_2O_3 Nps at comparatively low temperature in a minimum number of steps, and formed particles will be reliable for developing an effective sensor for harmful pollutants. The present study has emphasized a hydrothermal route for the preparation of Gd_2O_3 Nps under mild conditions. The hydrothermal process is the best preference due to its superior competence, economical nature, flexible reaction constraints, and prospective ability for the large-scale production of particles. The used methodology has provided better control over the size and shape of the formed particles. Abdullah et al. [29] have reported the fabrication of well crystalline Gd_2O_3 nanostructures by annealing the hydrothermally prepared nanostructures at 1000 °C. The prepared particles were further used for the detection of ethanol. In the current study, the biocompatible coating of ethylene glycol (EG) has provided better control over the agglomeration rate of formed nanoparticles. The presence of EG has a direct influence over the solubility, optical, luminescence, and the morphological characteristics of the prepared nanoparticles. The external template of EG has also modulated the range of non-radiative energy losses in Gd_2O particles.

Here in this work, surface-modified gadolinium oxide (Gd_2O_3) nanoparticles have been used as a proficient electrocatalytic material for the detection of hydrazine and p-nitrophenol by using cyclic voltammetric (CV) and chronoampherometric methods at neutral pH range. The consequences of synthetic parameters such as the concentration of precursors were studied by measuring the optical, photoluminescence, and band-gap variation of the formed particles. The estimation of the sensitivity, selectivity, repeatability, recyclability, linear range, detection limit, and applicability in real water has also been carried out in the current work. The thermal stability, resistance toward corrosion, and decreased tendency toward photobleaching made Gd_2O_3 nanoparticles a probable contender for preparing a simple, fast, and economical electrochemical sensor for hydrazine and p-nitrophenol.

2. Experimental Details

2.1. Materials

$GdCl_3 \cdot 6H_2O$ (Gadolinium(III) chloride hexahydrate: Sigma Aldrich, Mumbai, India with purity 99%) was used as a starting material for fabricating Gd_2O_3 nanoparticles. EG (ethylene glycol: Fluka 98%) and NaOH (sodium hydroxide: Merck 99.9% pure) were used for the synthesis purpose. Hydrazine, p-nitrophenol, benzaldehyde, benzoic acid, benzonitrile, phenol, ethanol, and aniline were procured from Sigma Aldrich, with purity more than 90%. Acetone (BDH, Mumbai, India, 98%) and ethanol (Changshu Yangyuan, Suzhou, China, 99.9%) were used as the washing solvent for obtained nanoparticles. Millipore distilled water was used for the synthesis of nanoparticles.

2.2. Synthesis of Gd_2O_3 Nanoparticles

The synthesis of Gd_2O_3 nanoparticles was done by using the hydrothermal method. In brief, 5 mM of $GdCl_3 \cdot 6H_2O$ was added to the 5-mL EG solution under stirring at 50 °C solution followed by the addition of 15 mM NaOH. The temperature of the reaction mixture was held constant at 140 °C for the first hour, and then raised to 180 °C for 4 h. The obtained solution was allowed to cool at room temperature. The obtained yellow precipitates of Gd_2O_3 nanoparticles were separated out from the reaction media. The obtained precipitates were subjected to calcinations for 3 h at 300 °C. The resulting particles were washed with water, acetone, and ethanol to remove the impurities. The corresponding separation was mainly performed by ultracentrifugation at 9000 rpm. The extracted particles were dried in an oven at 50 °C and further utilized for different analysis. For the optimization of synthetic parameters for the preparation of Gd_2O_3 nanoparticles, the respective concentration variations of $GdCl_3 \cdot 6H_2O$ have been carried out under respective reaction conditions. In the first instance, the concentration of variations of $GdCl_3 \cdot 6H_2O$ were done from 5 to 25 mM in all the reaction mixtures by keeping the concentration of the NaOH fixed at 0.015 M. A UV-visible spectral scan for each sample was taken from 230 to 400-nm wavelength to detect the optical properties of the formed nanoparticles. The optical band gap (Eg) of as-prepared nanoparticles was calculated as a function of the concentration variations of $GdCl_3 \cdot 6H_2O$. The respective fluorescence emission spectra were also studied for the concentration variations of $GdCl_3 \cdot 6H_2O$ from 5 to 25 mM.

2.3. Electrode Preparation

The synthesized nanoparticles were further used to fabricate the electrochemical sensor for hydrazine and p-nitrophenol (Scheme 1). In order to form the modified electrode, the gold electrode with a surface area equivalent to 3.14 mm^2 was first cleaned with alumina slurry and properly washed with distilled water under sonication. After drying the electrode at room temperature, the surface of the gold electrode was coated with the Gd_2O_3 nanoparticles by using butyl carbitol acetate (BCA) as the binding agent. The as-formed electrode was further dried at 60 ± 5 °C for 4–6 h to attain a homogeneous and dried layer of nanoparticles over the surface of electrode. All the electrochemical measurements were carried out on an μAutolab Type-III under neutral pH conditions. In all the analyses, a Gd_2O_3 customized gold electrode was acting as a working electrode, Ag/AgCl (sat. KCl) was acting as a reference, and Pt wire was acting as a counter electrode.

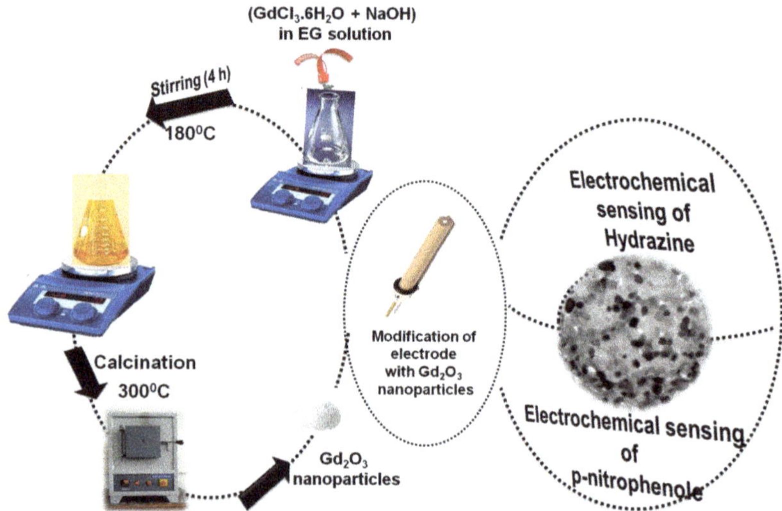

Scheme 1. Schematic illustration of the electrochemical sensor for the detection of hydrazine and p-nitrophenol by using the modified electrode of gold with Gd_2O_3 nanoparticles.

2.4. Physical Measurements

The obtained Gd_2O_3 nanoparticles were characterized with the help of an X-ray diffractometer from Panalytical D/Max-2500 (Malvern, UK), Hitachi (H-7500) Transmission electron microscope (Tokyo, Japan), Thermoscientific UV-vis. Spectrophotometer (Waltham, MA, USA), and FTIR spectrophotometer of Perkin-Elmer (RX1) (Waltham, MA, USA). The fluorescence measurements were carried out with a Hitachi F-7000 photoluminescence spectrophotometer. The photoluminescence analysis was carried out on an Edinburgh Instrument FLS 980 (Bain Square, UK). A JEOL (JSM-6610) scanning electron microscope (SEM) (Tokyo, Japan) with EDX analysis was carried out at 20 kV. The calcinations of the as-prepared Gd_2O_3 nanoparticles were done in an AICIL muffle furnace at 300 °C. Dynamic light scattering measurements were done on a Malvern Zen1690 instrument (Worcestershire, UK). Raman analysis was performed on a Renishaw inVia reflex micro-Raman spectrometer (Wotton-under-Edge, UK). The pH measurements were performed on a Mettler Toledo digital pH meter (Columbus, OH, USA). The surface area of EG-coated Gd_2O_3 nanoparticles was estimated by using Brunauer–Emmett–Teller (BET) analysis with an N_2 adsorption analyzer (NOVA 2000e, Anton Par, Gurugram, India). The separation of the as-prepared Gd_2O_3 nanoparticles from aqueous media was done on a Remi R-24 centrifuge. Electrochemical measurements were carried out on an µAutolab Type-III cyclic voltammeter (Metrohm, Herisau, Switzerland). The gold electrode with a surface area of 3.14 mm^2 was chosen for the analysis.

3. Results and Discussion

3.1. Characterization and Properties of Synthesized Gd_2O_3 Nanoparticles

The crystal structure of formed Gd_2O_3 nanoparticles has been further scrutinized by investigating the powdered XRD patterns of formed particles (Figure 1a). The absence of any peak related to impurity has confirmed the purity of the prepared nanoparticles [30]. The average crystallite size (D) of 15 nm has been estimated from the diffraction peaks by using the respective values of full width at half maximum (FWHM) via employing Debye–Scherrer's formula [31,32]. The specific surface area and pore diameter of the obtained sample was found to be 15.3 m^2·g^{-1} and 2.3 nm, which were respectively calculated by using the nitrogen sorption studies at 77 K in accordance with the BET (Brunauer–Emmett–Teller) process. The respective atomic content of Gd_2O_3 nanoparticles has been confirmed by using EDX analysis (Figure 1b). The obtained spectrum has only displayed the characteristic peaks of the Gd and O atoms in the synthesized sample, which verified the purity of the formed particles.

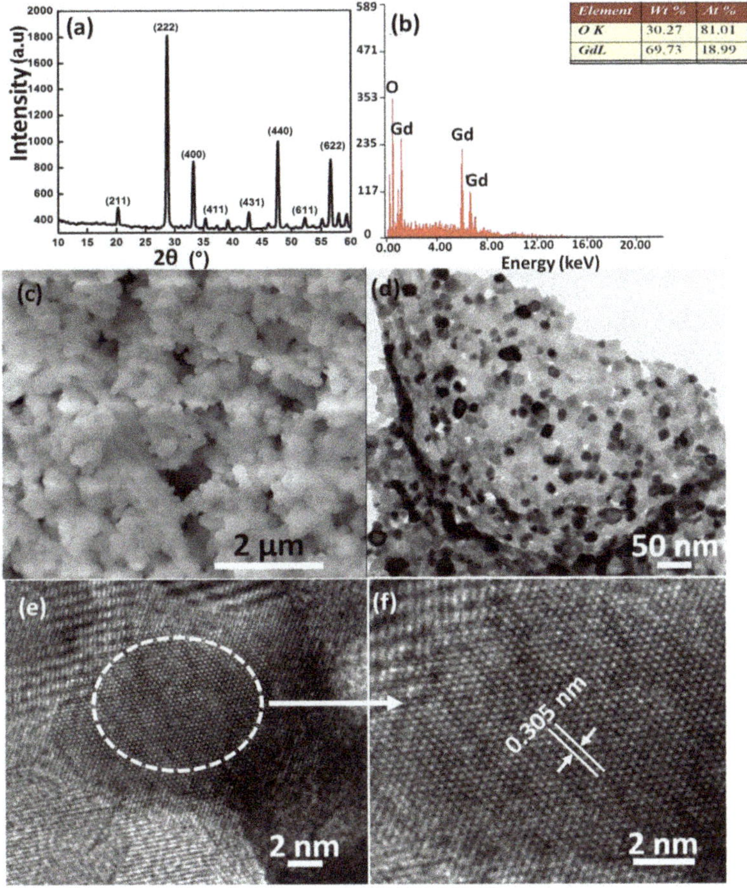

Figure 1. Typical (**a**) XRD pattern, (**b**) EDS spectrum, (**c**) SEM image, (**d**) TEM image, and (**e**,**f**) HRTEM images of Gd_2O_3 nanoparticles.

The SEM image of the Gd_2O_3 nanoparticles has been shown in Figure 1c, which has clearly shown the presence of agglomerated nanostructures of Gd_2O_3 nanoparticles. The presence of contacted particles has been mainly due to the existence of an EG template over the surface of the particles, which has further supported the presence of external electrostatic interactive forces generated from the templates over the surface of the nanoparticles. Detailed information regarding the morphology and structure of Gd_2O_3 nanoparticles has further been obtained from the HRTEM images presented in Figure 1d. It is clear from the images that the nanoparticles have shown crystallites with an irregular pseudo-spherical shape and a size distribution between 7–15 nm. The crystal spacing of 0.305 nm belongs to the (222) planes of Gd_2O_3 nanoparticles (Figure 1e,f). The observed result is in good agreement with the reported literature [33]. The presence of the connected nanocrystals has been attained due to the existence of a diverse range of forces (electrostatic, hydrogen bonding, and van der Waals forces) provided by the presence of external templates of EG coating over Gd_2O_3 nanoparticles.

The optical properties of formed Gd_2O_3 nanoparticles have been investigated by using UV-vis. and fluorescence analysis as a function of variation of the concentration of $GdCl_3 \cdot 6H_2O$ salt during the synthesis (Figure 2a,b). The formed particles have shown the characteristic peak between 255–262 nm. The respective peak has been associated with the electronic transition from $^8S_{7/2}$–$^6I_{7/2}$. [34]. On interpreting the results, it has been found that the absorbance is dependent on the concentration of salt.

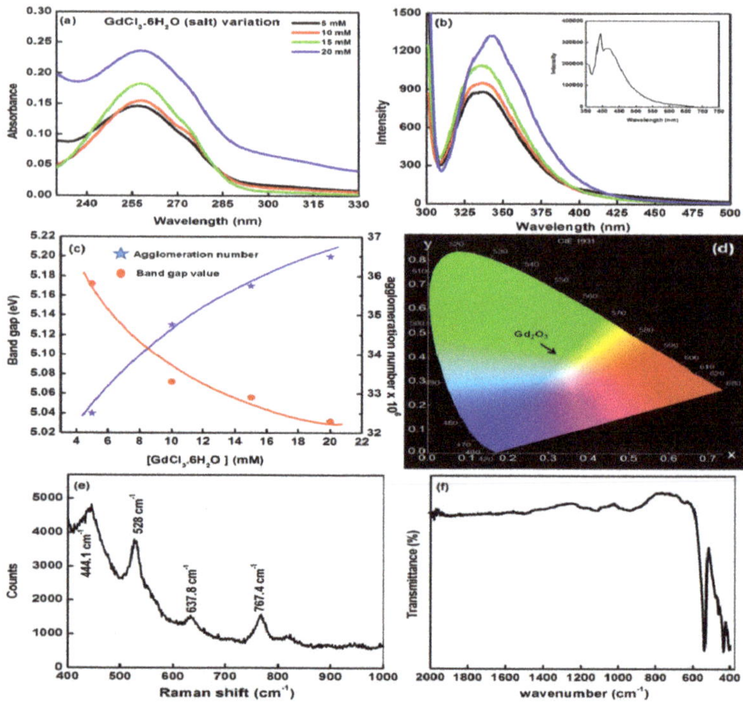

Figure 2. (**a**) UV-vis., (**b**) fluorescence and photoluminescence (PL) (inset) emission, (**c**) variation of bandgap and agglomeration number, (**d**) Commission Internationale de L'Eclairage (CIE) chromaticity analysis (**e**) Raman and (**f**) FTIR spectra of Gd_2O_3 nanoparticles.

With increase in the concentration from 5 to 20 mM, there has an increment in the absorbance of around 63%. The change in the peak position was not so prominent with the concentration of the salt. On the other hand, a fluorescence emission peak was observed at 336 nm with $\lambda_{exc} = 290$ nm (Figure 2b). This peak has been associated with the emission from $^6P_{7/2} \leftrightarrow ^8S_{7/2}$ in Gd(III) ions [35]. The enhancement in the concentration of the starting material has produced an increment of 50.7% in intensity value, whereas the peak position has only shown a change of 10 nm. This variation has displayed the similarity with UV-vis studies. The PL spectra of as-synthesized nanoparticles has shown the characteristic peaks at 390 nm and a broad peak between 400–500 nm with a center at 417 nm ($\lambda_{exc} = 350$ nm) (inset Figure 2b). The sharp peak at 390 nm has been associated with the radiative recombination of holes with electrons present at the oxygen vacant positions formed due to the photogeneration effect. The other peak was associated with the self-trapped exciton luminescence in formed particles [36]. The optical band-gap values (E_g) and agglomeration number for Gd_2O_3 nanoparticles (Figure 2c) have been estimated by using the application of the Brus method [37]. The results have clearly explained the behavioral variation of band-gap value with the concentration of $GdCl_3 \cdot 6H_2O$ salt during the synthesis. The decrease in the value of the band gap with the concentration has been associated with the variation of size of the formed particles with the concentration of the starting material. The particle size was comparatively higher for the nanoparticles prepared with 25 mM GdCl$_3$·6H$_2$O salt. These variations of starting material have directly influenced the size of the particles, and the respective agglomeration number of the particles has also varied in a similar manner. Figure 2d has displayed the respective assignment of their colors in the Commission Internationale de L'Eclairage (CIE) diagram for the Gd$_2$O$_3$ nanoparticles. The respective value of CIE chromaticity coordinates is found to be $x = 0.3265$, $y = 0.4462$, respectively. The outcomes have been associated with the green-shift effect in the formed particles. The formed particles have been further characterized by using the Raman and FTIR spectra of Gd$_2$O$_3$ nanoparticles (Figure 2e,f). On interpretation, Gd$_2$O$_3$ nanoparticles displayed four major Raman peaks at 444.1, 528, 637.8 and 767.4 cm^{-1}, respectively (Figure 2e). These peaks are mainly associated with the Fg and Ag mode for cubic C-type Gd$_2$O$_3$ particles [38]. Moreover, other Raman active modes such as 4A$_g$, 4E$_g$, and 14F$_g$ have also contributed toward the peaks in the spectra [39,40]. A sharp IR peak has been detected below 500 cm^{-1}, which has been attributed to Gd–O [40]. The small peak at 1500 cm^{-1} has been associated with the δ(O–H) vibrations due to the water bound to the nanoparticles surface in the form of moisture [41–44].

3.2. Electrochemical Behaviour of Hydrazine and p-Nitrophenol on Modified Electrode

The electrochemical action of formed Gd$_2$O$_3$ nanoparticles has been investigated toward the electrocatalysis of hydrazine and p-nitrophenol (PNP) via using cyclic voltammetric analysis. Figure 3 showed the cyclic voltammograms of a gold electrode in pH 7.0, phosphate buffer (PBS) under various electrode conditions. Interestingly, it has been found that the bare gold electrode does not show any signal in pH 7 PBS buffer. The response has been still negligible for bulk Gd$_2$O$_3$-coated gold electrodes in the presence of hydrazine and p-nitrophenol (Figure 3). On the other hand, well-defined voltammetric signals at 0.68 V have been obtained for the electrocatalysis of hydrazine in the presence of a Gd$_2$O$_3$ nanoparticles-coated gold electrode. In case of p-nitrophenol, one set of reversible redox peaks i.e., (Gd$_{R1}$), oxidation (Gd$_{O1}$) occurred at −0.021 and 0.163 V. Other irreversible reduction peaks (Gd$_{R2}$) at −0.694 V and (Gd$_{R3}$) at 0.4 V were also observed in phosphate buffer solution at pH 7 with a scan rate of 60 mV/s. The obtained results have clearly pointed out that the p-nitrophenol has displayed three types of electrochemical responses with two subsequent types of processes, including reduction and redox couple progression [21]. This might be aroused due to the two-electron oxidation reduction reaction in 4-aminophenol. Whereas, the reduction peaks were associated with the formation of a hydroxylamine group from the nitro group of p-nitrophenol.

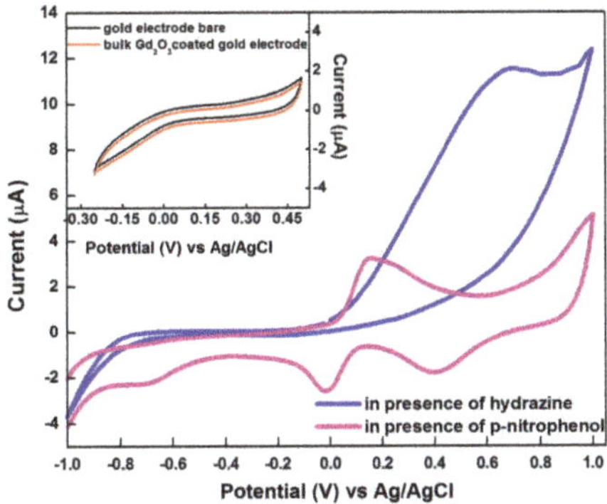

Figure 3. Cyclic voltammograms for a Gd$_2$O$_3$ nanoparticles-coated gold electrode under various electrode conditions in 0.1 M PBS (pH 7.0). The scan rate was 60 mV/s and the respective concentrations of analyte was kept constant at 1 mM.

From Figure 3, it was found that there has been no signal in the reverse sweep for hydrazine samples. The results have confirmed the irreversible nature of the oxidation process for hydrazine.

On other hand, the samples of p-nitrophenol have displayed the redox peaks in both forward and backward directions, which make the analysis of p-nitrophenol reversible in nature [45]. In the presence of Gd_2O_3 nanoparticles, there has been found a significant increase in the peak current for respective analytes. This has clearly verified the utilities of formed nanoparticles for electro-analytical purposes. The enhancement of peak current has been mainly explained by the enhancement of conductivity of the electrode due to the functionalization of the gold electrode surface with Gd_2O_3 nanoparticles. These results have further validated that the formed Gd_2O_3 nanoparticles are capable as efficient electron transporters for the electrocatalysis of harmful pollutants.

The scan rate variations have also been carried out in order to investigate the electron transfer mechanism for a Gd_2O_3 modified gold electrode in the presence of hydrazine and p-nitrophenol. Figure 4a,b shows the typical voltammograms for a Gd_2O_3 nanoparticles-coated gold electrode with 1 mM solution of respective analyte (i.e., hydrazine and p-nitrophenol) in 0.1 M PBS solution with pH = 7.0 at different scan rates. The spectrum has revealed a regular enhancement of current response with varying the scan rate from 60 to 900 mV/S for both the analytes. The higher surface area of Gd_2O_3 nanoparticles has played a critical role for the improved electron transfer process for the catalytic performance toward the understudied analytes. The enhancement of peak current with the scan rate has clearly pointed out the electrocatalytic reaction of hydrazine and p-nitrophenol at the surface of the modified electrode. These current variations have been mainly explained by the surface adsorption to diffusion processes at a façade of modified electrodes [45].

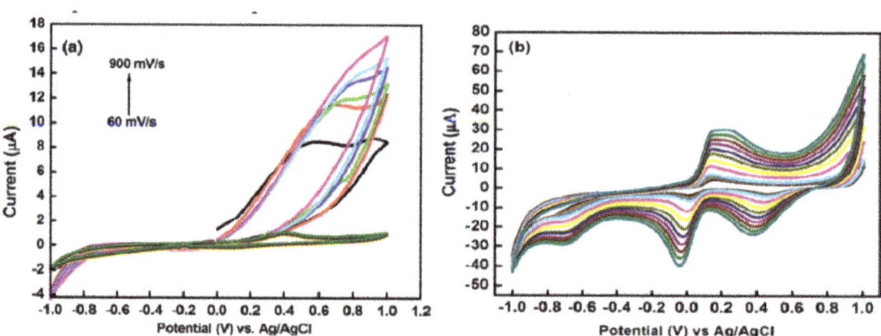

Figure 4. Cyclic voltammograms for a Gd_2O_3 nanoparticles-coated gold electrode for 1 mM (**a**) hydrazine and (**b**) p-nitrophenol at different scan rates ranging from 60 to 900 mV/s.

At lower scan rates, the relative rate of diffusion is quick, and the adsorption of analyte has been found to be slowest during the mass transfer process. On the other hand, at higher scan rates, the diffusion step has been mainly controlling the rate of electrode reactions in the presence of external agents [46]. In the case of p-nitrophenol, both types of associated processes i.e., the reduction and redox couple processes, have shown the significant augmentation of current response as a function of scan rate. The respective calibration curve of peak current versus the square root of the scan rate has displayed a linear relation for hydrazine and p-nitrophenol (Figure 5). These obtained results have clearly pointed out that the oxidation of hydrazine is mainly a diffusion-controlled process at the surface of a Gd_2O_3 nanoparticles-modified gold electrode [47].

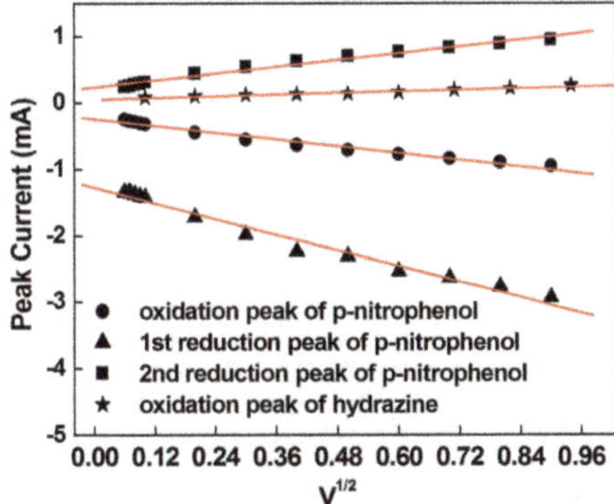

Figure 5. The linear dependence of peak current versus square root of scan rate for 1 mM hydrazine and p-nitrophenol.

The number of electrons involved in the overall reaction (n) for hydrazine and p-nitrophenol has been calculated from the Randles–Sevcik equation mentioned below [45].

$$ip = \left(2.69 \times 10^5\right) n^{\frac{3}{2}} A D^{\frac{1}{2}} v^{\frac{1}{2}} C \tag{1}$$

where n is the number of electron equivalents exchanged during the redox process, A (cm^2) is the active area of the working electrode, D (cm^2·s^{-1}) and C (mol·cm^{-3}) are the diffusion coefficient and the bulk concentration of hydrazine, and v is the voltage scan rate (V·s^{-1}). The obtained CV responses of Gd$_2$O$_3$/Au have clearly explained that the oxidation of N$_2$H$_4$ involves two electron changes, and the irreversible reaction of p-nitrophenol has undergone four electron changes, while the reversible reaction involves two processes—an electron redox process and an irreversible process—and gives a large reduction peak. (Figure 4b) shows the reduction of PNP to 4-(hydroxyamino) phenol. Two coupled redox peaks have indicated the oxidation of 4-(hydroxyamino) phenol to 4-nitrosophenol and the succeeding reversible reduction [48–51]. A schematic representation of the involved mechanism has been illustrated in Scheme 2.

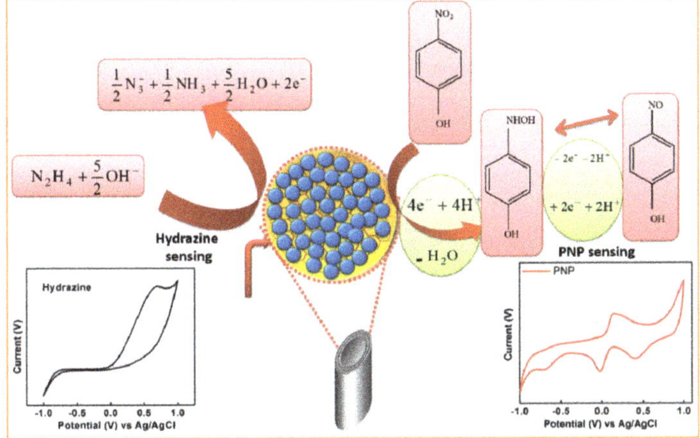

Scheme 2. The pictorial representation and cyclic voltammetric sweep curves for the Gd$_2$O$_3$ nanoparticles/butyl carbitol acetate/gold (NPs/BCA/Au) electrode for the sensing of hydrazine and p-nitrophenol.

3.3. Amperometric Responses for Hydrazine and p-Nitrophenol

The amperometric studies are one of the primary techniques to estimate the low concentration of analytes and carry out the relative studies in the presence of interfering analytes. Since the Gd_2O_3@Au-modified electrode has displayed the higher current response for hydrazine and p-nitrophenol as a model system in the cyclic voltammetric studies, therefore, it has been employed as the amperometric sensor for the detection of hydrazine and p-nitrophenol at low concentration levels.

Figures 6 and 7 depict the amperometric response of Gd_2O_3@Au for the successive additions of hydrazine and p-nitrophenol at an applied potential of −0.694 and 0.640 V for p-nitrophenol and hydrazine, respectively, in 0.1 M buffer solution with pH = 7. The obtained values of the current response have been estimated after the consecutive injection of 1 μM concentration of hydrazine and p-nitrophenol at the time interval of 60 s in a continuously stirring condition. The Gd_2O_3@Au-modified electrode has exhibited a considerable and rapid amperometric reaction toward each addition of analyte.

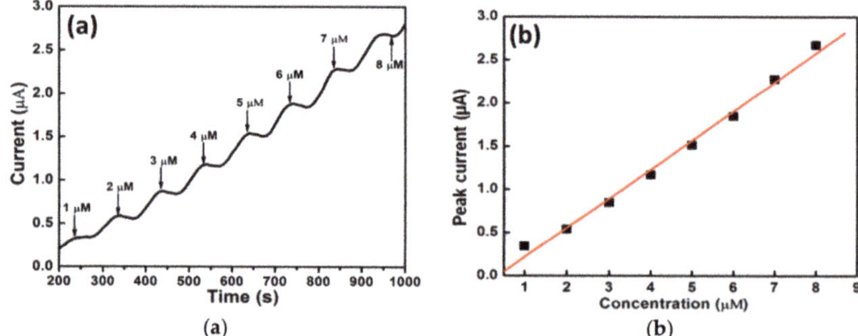

Figure 6. (**a**) Amperometric response of the Gd_2O_3/Au electrode with an increase in the concentration of hydrazine and (**b**) respective peak current vs. concentration plot of hydrazine in PBS at pH = 7.

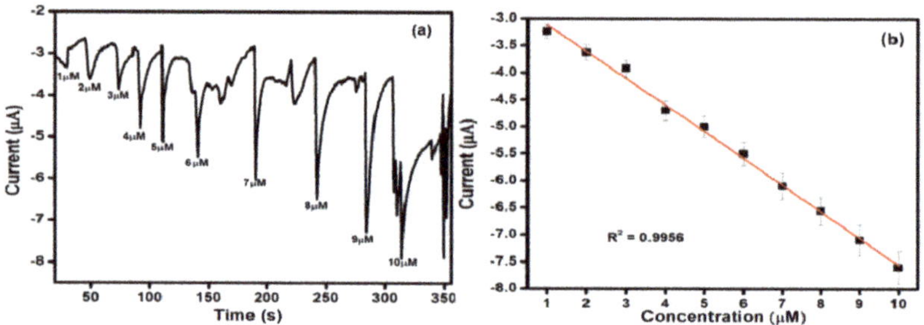

Figure 7. (**a**) Amperometric response of the Gd_2O_3/Au electrode with an increase in the concentration of p-nitrophenol and (**b**) respective peak current vs. concentration plot of p-nitrophenol in PBS at pH = 7.

The value of the current has reached its stable position within 3 s, demonstrating the fast electro-oxidation of the understudied analyte at the surface of the Gd_2O_3@Au-modified electrode. In addition, the response current has shown a linear increment for the subsequent additions of analyte over the wide range of concentrations. The respective regression plot of current response versus concentration of both the analytes has displayed a linear relation with the correlation coefficient values of 0.987 and 0.996 for p-nitrophenol and hydrazine, respectively (Figures 6b and 7b). The limit of detection value has been calculated to be 1.527 and 0.704 μM for p-nitrophenol and hydrazine, respectively, by using the equation limit of detection (LOD) = 3σ/slope, where σ is the standard deviation for the particular system [52]. The sensitivity of the developed sensor has been found to be 0.33722 and 0.25734 mA·mM^{-1} from the slope of the linear regression for hydrazine and p-nitrophenol, respectively. The respective reusability, stability, and reproducibility of the as-prepared sensor has also been tested in the current work. The as-developed electrode was kept in the buffer media for one month, and its electrocatalytic efficiency has also been tested against the p-nitrophenol and hydrazine. The results have clearly verified that the formed sensor has displayed a reproducible performance with a decay rate of 5.7% and 6.3% in the oxidation peak current value for p-nitrophenol and hydrazine, respectively. This substantial constancy in results has been further attributed to the stability of Gd_2O_3 particles, which maintain the efficiency and performance of the electrode for a long period. Additionally, the reproducibility of the developed sensor has been confirmed by estimating the electrochemical response of the Gd_2O_3 particles as a working electrode for different electrodes in the solution media containing 1 mM of p-nitrophenol and hydrazine.

The obtained relative standard deviation (RSD) of peak currents has been found to be 5.3% and 4.7%, signifying the satisfactory report for the reproducibility of the modified electrode. In order to investigate the reusability of the developed sensor, the as-modified electrode has been rinsed with the respective sample solution. The obtained signal has been tested after the rinsing. It has been found that the obtained signal has maintained 94% of its original strength. The analytical performance of the as-modified electrode has been evaluated with some recent works in Table 1.

Table 1. Comparison of detection limit and response time of different electrode materials.

Electrode Materials	Analyte	Detection Limit/μM	Response Time/s	Refs.
Copper tetraphenylporphyrin (CuTPP) onto zeolites cavity-modified carbon paste electrode	Hydrazine, p-nitrophenol	1	–	[53]
Multi wall carbon nanotubes (MWCNT) and chlorogenic acid	Hydrazine	8	–	[54]
Single wall carbon nanotube (SWCNT) and catechin hydrate	Hydrazine and hydroxyl amine	2.0	–	[54]
Nickel hexacyanoferrate modified carbon ceramic electrode	Hydrazine	2.28	<3	[55]
Carbon nanotubes powder microelectrode	Hydrazine	–	<3	[56]
ZnO nanorods	Hydrazine, p-nitrophenol	2.2	<10	[57]
Gd_2O_3 nanoparticles	–	0.704	<10	This work

The data has further confirmed the authenticity of the developed sensor in a different range of concentrations with greater selectivity and sensitivity. In order to investigate the application of the formed sensor in real samples, the water samples from different sources have been taken, and the respective analyses have been made for the detection of hydrazine and p-nitrophenol. The relevant stock solutions of understudied analytes of known concentrations were made by using the water samples taken from different sources. The current response of the sensor has been examined for the different concentrations of hydrazine and p-nitrophenol by using amperometric studies. The outcomes of the measurements have shown the excellent recovery rate for the chosen analytes, as indicated in Table 2.

The sensitivity, selectivity, repeatability, recyclability, wide linear range, detection limit, and applicability in real water samples makes Gd_2O_3 Nps a favorable nanomaterial in the institution of the rapid and effectual scrutiny of harmful environmental pollutants. Thus, the prepared sensor appears to be a probable contender for preparing a simple, fast, and economical electrochemical sensor.

Table 2. Determination of hydrazine and p-nitrophenol real water samples. RSD: relative standard deviation.

Sample	Added Amount (μM)	Hydrazine	p-Nitrophenol
		Recovery Mean ± RSD (%)	
Tap water	1.5	98.2 ± 2.9	97.6 ± 2.5
	3.5	100.2 ± 1.8	99.7 ± 3.9
	7.5	101.9 ± 1.5	100.3 ± 3.7
Lake water	1.5	96.6 ± 1.5	97.3 ± 1.1
	3.5	96.9 ± 2.6	99.1 ± 2.4
	7.5	98.4 ± 3.5	100.1 ± 1.6
Water from Village Dhanas	1.5	99.6 ± 2.3	97.4 ± 1.3
	3.5	100.2 ± 1.4	99.4 ± 2.8
	7.5	101.5 ± 1.2	100.2 ± 1.2

3.4. Selectivity Study

In order to apply the proposed sensor for the determination of p-nitrophenol and hydrazine in an aqueous system, the respective selectivity of the sensor has been investigated in the presence of 1 mM of various interfering compounds (benzaldehyde, benzoic acid, benzonitrile, phenol, ethanol, and aniline) at a fixed potential 0.67 V for hydrazine and −0.69 for PNP. Figure 8 has displayed the amperometric response of a Gd_2O_3/BCA/Au electrode for the same. From the data, it has been found that a negligible change to response current has been detected in the presence of different interfering compounds. However, significant and quick responses were observed in the presence of hydrazine and PNP. These outcomes have clearly explained the high selectivity of the fabricated sensor toward hydrazine and PNP, and enhanced the scope of the developed sensor.

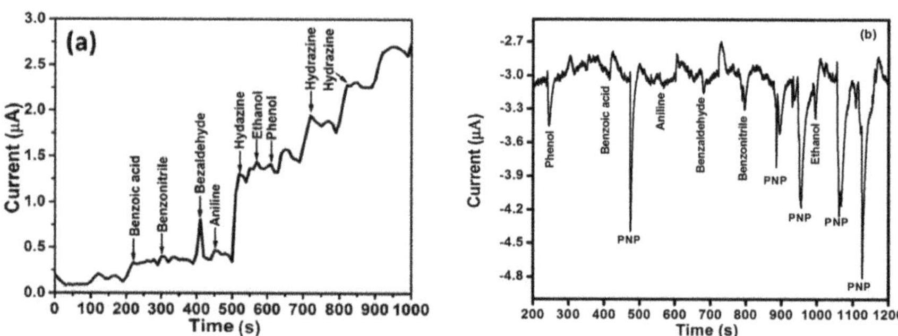

Figure 8. Amperometric response of the Gd_2O_3/Au electrode in the presence of different analytes with a concentration of 1 mM at pH 7 for (**a**) hydrazine and (**b**) PNP.

4. Conclusions

In summary, the current work has reported the fabrication of the ethylene glycol-mediated synthesis of Gd_2O_3 Nps. The formation of the particles has been scrutinized by using sophisticated characterization techniques. The effects of synthetic parameters over the optical, photoluminescence, band-gap variation, and agglomeration number of the formed particles have been studied in detail. The fabricated particles have further been employed as a proficient electrocatalytic material for the enzyme-free detection of hydrazine and p-nitrophenol with great sensitivity and selectivity. The developed sensors hold a wider linear range of 1 to 10 μM with low detection limits of 1.527 μM and 0.704 μM for p-nitrophenol and hydrazine, respectively. The sensitivity, selectivity, repeatability, recyclability, wide linear range, and detection limit make Gd_2O_3 Nps a favorable nanomaterial in the institution of the rapid and effectual scrutiny of harmful environmental pollutants. The realistic application of the developed sensor has also been investigated by spiking known concentrations of hydrazine and p-nitrophenol in different water samples with good recoveries. Consequently, the successful synthesis of EG@Gd$_2$O$_3$ nanoparticles has immense potential for the design of highly effective electrochemical sensors, and is a probable way to provide momentum to the advancement of new electrode materials.

Author Contributions: S.C. for methodology, validation, writing; S.K. (Sandeep Kumar) and S.K. (Sushil Kumar) for formal analysis; A.U., G.R.C. and S.K.M. for data curation.

Funding: This research was funded by DST Inspire Faculty award [IFA-CH-17] and DST purse grant II. Sandeep Kumar is grateful to CSIR India for providing a senior research fellowship. Ahmad Umar would like to acknowledge the Ministry of Education, Saudi Arabia for this research through a grant (PCSED-013-18) under the Promising Centre for Sensors and Electronic Devices (PCSED) at Najran University, Kingdom of Saudi Arabia.

Conflicts of Interest: The authors declare no conflict of interest.

References

1. Li, J.; Kuang, D.; Feng, Y.; Zhang, F.; Xu, Z.; Liu, M. A graphene oxide-based electrochemical sensor for sensitive determination of 4-nitrophenol. *J. Hazard Mater.* **2012**, *201*, 250–259. [CrossRef] [PubMed]

2. Nam, G.; Leem, J.-Y. A new technique for growing ZnO nanorods over large surface areas using graphene oxide and their application in ultraviolet sensors. *Sci. Adv. Mater.* **2018**, *10*, 405–409. [CrossRef]

3. Vernot, E.H.; MacEwen, J.D.; Bruner, R.H.; Haus, C.C.; Kinkead, E.R.; Prentice, D.E.; Hall, A., III; Schmidt, R.E.; Eason, R.L.; Hubbard, G.B.; et al. Long-term inhalation toxicity of hydrazine. *Fundam. Appl. Toxicol.* **1985**, *5*, 1050–1064. [CrossRef]

4. Savafi, A.; Karimi, M.A. Flow injection chemiluminescence determination of hydrazine by oxidation with chlorinated isocyanurates. *Talanta* **2002**, *58*, 785–792. [CrossRef]

5. Safavi, A.; Ensafi, A.A. Electrocatalytic oxidation of hydrazine with pyrogallol red as a mediator on glassy carbon electrode. *Anal. Chem. Acta* **1995**, *300*, 307–311. [CrossRef]

6. Chen, H.; Li, D.; Ding, W. Template-less preparation of meso-MnO$_2$ fibers by localized ostwald ripening and 2,4-dinitrophenol adsorption mechanism. *Sci. Adv. Mater.* **2018**, *10*, 1241–1249. [CrossRef]

7. U.S. Department of Homeland Security. *Homeland Security Information Bulletin*; U.S. Department of Homeland Security: Washington, DC, USA, 2003.

8. Shukla, S.; Chaudhary, S.; Umar, A.; Chaudhary, G.R.; Mehta, S.K. Dodecyl ethyl dimethyl ammonium bromide capped WO$_3$ nanoparticles: Efficient scaffolds for chemical sensing and environmental remediation. *Dalton Trans.* **2015**, *44*, 17251–17260. [CrossRef]

9. Collins, G.E.; Rose-Pehrsson, S.L. Sensitive, fluorescent detection of hydrazine via derivatization with 2,3-naphthalene dicarboxaldehyde. *Analytica Chimica Acta* **1993**, *284*, 207–215. [CrossRef]

10. Liu, X.; Yang, Z.; Sheng, Q.; Zheng, J. One-pot synthesis of Au-Fe$_3$O$_4$-GO nanocomposites for enhanced electrochemical sensing of hydrazine. *J. Electrochem. Soc.* **2018**, *165*, B596–B602. [CrossRef]

11. Gao, Y.; Zhang, S.; Hou, W.; Guo, H.; Li, Q.; Dong, D.; Wu, S.; Zhao, S.; Zhang, H. Perylene diimide derivative regulate the antimony sulfide morphology and electrochemical sensing for hydrazine. *Appl. Surf. Sci.* **2019**, *491*, 267–275. [CrossRef]

12. Guo, W.; Xia, T.; Zhang, H.; Zhao, M.; Wang, L.; Pei, M.A. Molecularly imprinting electrosensor based on the novel nanocomposite for the detection of tryptamine. *Sci. Adv. Mater.* **2018**, *10*, 1805–1812. [CrossRef]

13. Annalakshmi, M.; Balasubramanian, P.; Chen, S.-M.; Chen, T.-W. One pot synthesis of nanospheres-like trimetallic NiFeCo nanoalloy: A superior electrocatalyst for electrochemical sensing of hydrazine in water bodies. *Sens. Actuators B* **2019**, *296*, 126620. [CrossRef]

14. Gu, D.; Dong, Y.-M.; Liu, H.; Ding, H.-Y.; Shang, S.-M. Efficient synthesis of highly fluorescence nitrogen doped carbon dots and its application for sensor and cell imaging. *Sci. Adv. Mater.* **2018**, *10*, 964–973. [CrossRef]

15. Mishra, R.K.; Martin, A.; Nakagawa, T.; Barfidokht, A.; Lu, X.; Sempionatto, J.R.; Lyu, K.M.; Karajic, A.; Musameh, M.M.; Kyratzis, I.L.; et al. Detection of vapor-phase organophosphate threats using wearable conformable integrated epidermal and textile wireless biosensor systems. *Biosens. Bioelectron.* **2018**, *101*, 227–234. [CrossRef] [PubMed]

16. Mishra, R.K.; Hubble, L.J.; Martin, A.; Kumar, R.; Barfidokht, A.; Kim, J.; Musameh, M.M.; Kyratzis, I.L.; Wang, J. Wearable flexible and stretchable glove biosensor for on-site detection of organophosphorus chemical threats. *ACS Sen.* **2017**, *2*, 553–561. [CrossRef] [PubMed]

17. Wei, Y.; Kong, L.T.; Yang, R.; Wang, L.; Liu, H.; Huang, X.J. Single-walled carbon nanotube/pyrene cyclodextrin nanohybrids for ultrahighly sensitive and selective detection of *p*-nitrophenol. *Langmuir* **2011**, *27*, 10295–10301. [CrossRef]

18. Karthik, R.; Chen, S.M.; Elangovan, A.; Muthukrishnan, P.; Shanmugam, R.; Lou, B.S. Phyto mediated biogenic synthesis of gold nanoparticles using *Cerasus serrulata* and its utility in detecting hydrazine, microbial activity and DFT studies. *J. Colloid Int. Sci.* **2016**, *468*, 163–175. [CrossRef]

19. Zhang, R.; Sun, C.L.; Lu, Y.J.; Chen, W. Graphene nanoribbon-supported Pt-Pd concave nanocubes for electrochemical detection of TNT with high sensitivity and selectivity. *Anal. Chem.* **2015**, *87*, 12262–12269. [CrossRef]

20. Chen, X.; Cheng, X.; Gooding, J.J. Detection of trace nitroaromatic isomers using Indium Tin oxide electrodes modified using β-cyclodextrin and silver nanoparticles. *Anal. Chem.* **2012**, *84*, 8557–8563. [CrossRef]

21. Chaudhary, S.; Kumar, S.; Mehta, S.K. Glycol modified gadolinium oxide nanoparticles as a potential template for selective and sensitive detection of 4-nitrophenol. *J. Mater. Chem. C* **2015**, *3*, 8824–8833. [CrossRef]

22. Du, G.; Tendeloo, G.V. Preparation and structure analysis of $Gd(OH)_3$ nanorods. *Nanotechnology* **2015**, *16*, 595–597. [CrossRef]

23. Mehta, S.K.; Chaudhary, S.; Bhasin, K.K. Understanding the role of hexadecyltrimethylammonium bromide in the preparation of selenium nanoparticles: A spectroscopic approach. *J. Nanopart. Res.* **2009**, *11*, 1759–1766. [CrossRef]

24. Ballem, M.A.; Söderlind, F.; Nordblad, P.; Kall, P.O.; Oden, M. Growth of Gd_2O_3 nanoparticles inside mesoporous silica frameworks. *Microporous Mesoporous Mater.* **2013**, *168*, 221–224. [CrossRef]

25. Rahman, A.T.M.; Vasilev, K.; Majewski, P. Ultra small Gd_2O_3 nanoparticles: Absorption and emission properties. *J. Colloid Inter. Sci.* **2011**, *354*, 592–596. [CrossRef]

26. Jia, G.; You, H.P.; Liu, K.; Zheng, Y.H.; Guo, N.; Zhang, H.J. Highly uniform Gd_2O_3 hollow microspheres: Template-directed synthesis and luminescence properties. *Langmuir* **2010**, *26*, 5122–5128. [CrossRef]

27. Hu, L.; Ma, R.; Ozawa, T.C.; Sasaki, T. Oriented monolayer film of Gd_2O_3:0.05Eu crystallites: Quasi-topotactic transformation of the hydroxide film and drastic enhancement of photoluminescence properties. *Angew. Chem. Int. Ed.* **2009**, *48*, 3846–3849. [CrossRef] [PubMed]

28. Manigandan, R.; Giribabu, K.; Suresh, R.; Vijayalakshmi, L.; Stephen, A.; Narayanan, V. Structural, optical and magnetic properties of gadolinium sesquioxide nanobars synthesized via thermal decomposition of gadolinium oxalate. *Mater. Res. Bull.* **2013**, *48*, 4210–4215. [CrossRef]

29. Abdullah, M.M.; Rahman, M.M.; Bouzid, H.; Faisal, M.; Khan, S.B.; Al-Sayari, S.A.; Ismail, A.A. Sensitive and fast response ethanol chemical sensor based on as-grown Gd_2O_3 nanostructures. *J. Rare Earths* **2015**, *33*, 214–220. [CrossRef]

30. Chaudhary, S.; Kumar, S.; Umar, A.; Singh, J.; Rawat, M.; Mehta, S.K. Europium-doped gadolinium oxide nanoparticles: A potential photoluminescencent probe for highly selective and sensitive detection of Fe^{3+} and Cr^{3+} ions. *Sens. Actuators B* **2017**, *243*, 579–588. [CrossRef]

31. Mehta, S.K.; Chaudhary, S.; Kumar, S.; Bhasin, K.K.; Torigoe, K.; Sakai, H.; Abe, M. Surfactant assisted synthesis and spectroscopic characterization of selenium nanoparticles in ambient condition. *Nanotechnology* **2008**, *19*, 295601–295612. [CrossRef]

32. Chaudhary, S.; Sharma, P.; Kumar, R.; Mehta, S.K. Nanoscale surface designing of cerium oxide nanoparticles for controlling growth, stability, optical and thermal properties. *Ceram. Int.* **2015**, *41*, 10995–11003. [CrossRef]

33. Maalej, N.M.; Qurashi1, A.; Assadi, A.A.; Maalej, R.; Shaikh, M.N.; Ilyas, M.; Gondal, M.A. Synthesis of Gd_2O_3:Eu nanoplatelets for MRI and fluorescence imaging. *Nanoscale Res. Lett.* **2015**, *10*, 215. [CrossRef] [PubMed]

34. Mukherjee, S.; Dasgupta, P.; Jana, P.K. Size-dependent dielectric behaviour of magnetic Gd_2O_3 nanocrystals dispersed in a silica matrix. *J. Phys. D Appl. Phys.* **2008**, *41*, 215004–215015. [CrossRef]

35. Rogow, D.L.; Swanson, C.H.; Oliver, A.G.; Oliver, S.R.J. Two related Gadolinium aquo carbonate 2-D and 3-D structures and their thermal, spectroscopic, and paramagnetic properties. *Inorg. Chem.* **2009**, *48*, 1533–1541. [CrossRef] [PubMed]

36. Zhang, Y.; Han, K.; Cheng, T.; Fang, Z. Synthesis, characterization, and photoluminescence property of $LaCO_3OH$ microspheres. *Inorg. Chem.* **2007**, *46*, 4713–4717. [CrossRef] [PubMed]

37. Chaudhary, S.; Umar, A. Glycols functionalized fluorescent Eu_2O_3 nanoparticles: Functionalization effect on the structural and optical properties. *J. Alloys Compd.* **2016**, *682*, 160–169. [CrossRef]

38. Rajan, G.; Gopchandran, K.G. Enhanced luminescence from spontaneously ordered Gd_2O_3: Eu^{3+} based nanostructures. *Appl. Surf. Sci.* **2009**, *255*, 9112–9123. [CrossRef]

39. White, W.B.; Keramidas, V.G. Vibrational spectra of oxides with the C-type rare earth oxide structure. *Spectrochim. Acta Part A* **1972**, *28*, 501–509. [CrossRef]

40. Riri, M.; Benjjar, A.; Eljaddi, T.; Sefiani, N.; Touaj, K.; Cherif, A.; Hlaïbi, M. Characterization of two dinuclear complexes of the gadolinium ion by IR and Raman. *J. Mater. Environ. Sci.* **2013**, *4*, 961–966.

41. Coats, A.W.; Redfern, J.P. Kinetic parameters from thermogravimetric data. *Nature* **1964**, *201*, 68–69. [CrossRef]

42. Madhusudanan, P.M.; Krishnan, K.; Ninan, K.N. New approximation for the $p(x)$ function in the evaluation of non-isothermal kinetic data. *Thermochimica Acta* **1986**, *97*, 189–201. [CrossRef]

43. Tang, W.; Liu, Y.; Zhang, H.; Wang, Z.; Wang, C. New temperature integral approximate formula for non-isothermal kinetic analysis. *J. Therm. Anal. Calorim.* **2003**, *74*, 309–315. [CrossRef]

44. Horowitz, H.H.; Metzger, G.A. A new analysis of thermogravimetric traces. *Anal. Chem.* **1963**, *35*, 1464–1468. [CrossRef]

45. Singh, K.; Kaur, A.; Umar, A.; Chaudhary, G.R.; Singh, S.; Mehta, S.K. A comparison on the performance of zinc oxide and hematite nanoparticles for highly selective and sensitive detection of para-nitrophenol. *J. Appl. Electrochem.* **2015**, *45*, 253–261. [CrossRef]

46. Raoof, J.B.; Ojani, R.; Beitollahi, H.; Hosseinzadeh, R. Electrocatalytic oxidation and highly selective voltammetric determination of L-cysteine at the surface of a 1-[4-(ferrocenyl ethynyl)phenyl]-1-ethanone modified carbon paste electrode. *Anal. Sci.* **2006**, *22*, 1213–1220. [CrossRef] [PubMed]

47. Shahid, M.M.; Rameshkumar, P.; Basirunc, W.J.; Wijayantha, U.; Chiu, W.S.; Khiew, P.S.; Huang, N.M. An electrochemical sensing platform of cobalt oxide@gold nanocubes interleaved reduced graphene oxide for the selective determination of hydrazine. *Electrochimica Acta* **2018**, *259*, 606–616. [CrossRef]

48. Liu, Z.; Du, J.; Qiu, C.; Huang, L.; Ma, H.; Shen, D.; Ding, Y. Electrochemical sensor for detection of *p*-nitrophenol based on nanoporous gold. *Electrochem. Commun.* **2009**, *11*, 1365–1368. [CrossRef]

49. Ndlovu, T.; Arotiba, O.A.; Krause, R.W.; Mamba, B.B. Electrochemical detection of o-nitrophenol on a poly(propyleneimine)-gold nanocomposite modified glassy carbon electrode. *Int. J. Electrochem. Sci.* **2010**, *5*, 1179–1186.

50. Shukla, S.; Chaudhary, S.; Umar, A.; Chaudhary, G.R.; Mehta, S.K. Tungsten oxide (WO_3) nanoparticles as scaffold for the fabrication of hydrazine chemical sensor. *Sens. Actuators B* **2014**, *196*, 231–237. [CrossRef]

51. De Oliveira, F.M.; Guedesa, T.d.; Lima, A.B.; da Silvaa, L.M.; dos Santos, W.T.P. Alternative method to obtain the Tafel plot for simple electrode reactions using batch injection analysis coupled with multiple-pulse amperometric detection. *Electrochimica Acta* **2017**, *242*, 180–186. [CrossRef]

52. Wang, X.; Wang, Q.; Wang, Q.; Gao, F.; Gao, F.; Yang, Y.; Guo, H. Highly dispersible and stable copper terephthalate metal-organic framework-graphene oxide nanocomposite for an electrochemical sensing application. *ACS Appl. Mater. Interfaces* **2014**, *6*, 11573–11580. [CrossRef] [PubMed]

53. Guerra, S.V.; Ubota, L.T.K.; Xavier, C.R.; Nakagaki, S. Experimental optimization of selective hydrazine detection in flow injection analysis using a carbon paste electrode modified with copper porphyrin occluded into zeolite cavity. *Anal. Sci.* **1999**, *15*, 1231–1234. [CrossRef]

54. Salimi, A.; Miranzadeh, L.; Hallaj, R. Amperometric and voltammetric detection of hydrazine using glassy carbon electrodes modified with carbon nanotubes and catechol derivatives. *Talanta* **2008**, *75*, 147–156. [CrossRef] [PubMed]

55. Salami, A.; Abdi, K. Enhancement of the analytical properties and catalytic activity of a nickel hexacyanoferrate modified carbon ceramic electrode prepared by two-step sol–gel technique: Application to amperometric detection of hydrazine and hydroxyl amine. *Talanta* **2004**, *63*, 475–483. [CrossRef]

56. Zhao, Y.D.; Zhang, W.D.; Chen, H.; Luo, Q.M. Anodic oxidation of hydrazine at carbon nanotube powder microelectrode and its detection. *Talanta* **2002**, *58*, 529–534. [CrossRef]

57. Umar, A.; Rahman, M.M.; Hahn, Y.B. ZnO nanorods based hydrazine sensors. *J. Nanosci. Nanotechnol.* **2009**, *9*, 4686–4691. [CrossRef]

Article

Iron-Doped Titanium Dioxide Nanoparticles As Potential Scaffold for Hydrazine Chemical Sensor Applications

Ahmad Umar [1,2,*], Farid A. Harraz [2,3], Ahmed A. Ibrahim [1,2], Tubia Almas [1,2,4], Rajesh Kumar [5], M. S. Al-Assiri [2,6] and Sotirios Baskoutas [4]

[1] Department of Chemistry, Faculty of Science and Arts, Najran University, Najran-11001, Saudi Arabia; ahmedragal@yahoo.com (A.A.I.); tubiaalmas@gmail.com (T.A.)
[2] Promising Centre for Sensors and Electronic Devices (PCSED), Najran University, Najran-11001, Saudi Arabia; fharraz.68@yahoo.com (F.A.H.); msassiri@gmail.com (M.S.A.-A.)
[3] Nanostructured Materials and Nanotechnology Division, Central Metallurgical Research and Development Institute (CMRDI), P.O. Box: 87 Helwan, Cairo 11421, Egypt
[4] Department of Materials Science, University of Patras, GR-26504 Patras, Greece; bask@upatras.gr
[5] Department of Chemistry, Jagdish Chandra DAV College, Dasuya-144205, India; rk.ash2k7@gmail.com
[6] Department of Physics, Faculty of Science and Arts, Najran University, Najran-11001, Saudi Arabia
* Correspondence: umahmad@nu.edu.sa or ahmadumar786@gmail.com

Received: 5 January 2020; Accepted: 13 February 2020; Published: 17 February 2020

Abstract: Herein, we report the fabrication of a modified glassy carbon electrode (GCE) with high-performance hydrazine sensor based on Fe-doped TiO_2 nanoparticles prepared via a facile and low-cost hydrothermal method. The structural morphology, crystalline, crystallite size, vibrational and scattering properties were examined through different characterization techniques, including FESEM, XRD, FTIR, UV–Vis, Raman and photoluminescence spectroscopy. FESEM analysis revealed the high-density synthesis of Fe-doped TiO_2 nanoparticles with the average diameter of 25 ± 5 nm. The average crystallite size of the synthesized nanoparticles was found to be around 14 nm. As-fabricated hydrazine chemical sensors exhibited 1.44 µA µM^{-1} cm^{-2} and 0.236 µM sensitivity and limit of detection (LOD), respectively. Linear dynamic ranged from 0.2 to 30 µM concentrations. Furthermore, the Fe-doped TiO_2 modified GCE showed a negligible inference behavior towards ascorbic acid, uric acid, glucose, $SO_4{}^{2-}$, $NO_3{}^-$, Pb^{2+} and Ca^{2+} ions on the hydrazine sensing performance. Thus, Fe-doped TiO_2 modified GCE can be efficiently used as an economical, easy to fabricate and selective sensing of hydrazine and its derivatives.

Keywords: Fe-doped TiO_2; hydrothermal; GCE; hydrazine; chemical sensor; amperometry

---

1. Introduction

Hydrazine is extensively used in rocket propellants, pesticides, explosives, photography chemicals, antioxidants and plant growth regulators in various related industries and laboratories [1,2]. Effluents released from these industries and laboratories are the foremost environmental source of hydrazine. It is considered one of the poisonous chemicals even at low concentration and is very hazardous to living organisms. The potential symptoms of hydrazine exposure range from eye and nose irritation, pulmonary edema, skin dermatitis, temporary blindness and serious damage to many human organs such as the kidney and liver [3–5]. Due to its mutagenic and carcinogenic behavior, it has been cited in the Environmental Protection Agency (EPA) list. It is therefore mandatory to trace the presence of minor amounts of hydrazine in the aqueous medium.

Previously reported techniques for the detection of hydrazine include spectrophotometry [6], fluorimetry [7,8], chemiluminescence [9] and potentiometry [10,11]. However, all these methods are accompanied by several disadvantages like low sensitivity, and they require expensive instrumentation followed by complicated procedures. Electrochemical determination is considered to be the promising alternative for the determination of hydrazine in terms of selectivity, sensitivity and portability, with an economical and simple operating procedure [12,13]. Unfortunately, hydrazine undergoes direct oxidation at the bare electrode surface, thus resulting in sluggish electrode kinetics and high over potentials [14,15]. Therefore, chemically modified electrodes have been used to detect hydrazine, which significantly reduce the overpotential, as well as accelerate redox reactions and hence the oxidation current responses [16,17].

Recently, there is a big interest in the utilization of new materials in scientific societies due to improved processability, applicability and vast applications in various fields of sciences. In the past few years, tremendous work has been done on metal oxide based nanomaterials or composites, especially ZnO [18], CuO [19,20], TiO_2 [21,22] and MnO_2 [23], which are considered an ideal platform for material preparation used in pharmaceuticals and cosmetics industries, for the treatment of wastewater and in other fields [24]. TiO_2 is one of the most extensively used semiconductor materials owing to its non-toxicity, high performance, great stability and low preparation cost. Due to its good conductivity, high electrocatalytic activity and electron transport properties, TiO_2 is the most suitable material for devising the electrodes for various electrochemical and biosensing applications. Khodari et al. [25] used an electrochemical sensor, prepared by casting TiO_2 nanoparticles (NPs) onto a carbon electrode surface, for the determination of resorcinol. The TiO_2 NPs were demonstrated to be competent in boosting the electron transfer between resorcinol and the electrode surface as well as transport of resorcinol molecules to the surface of the sensor. Guo and coworkers [26,27] synthesized TiO_2 nanofibers for the bio-sensing of glucose. The fabricated TiO_2 nanofibers exhibited good direct electrochemistry as well as magnificent sensitivity along with fast response time for the detection of glucose. In order to further increase the electron transfer rate between the working modified electrode and the analyte solution and, hence, rapid and sensitive current response, doping of metal oxides has been reported as one of the best ways. Among the various dopants, Fe^{3+} ions are most suitable owing to their similar ionic radii to that of Ti^{4+} ions. Fe^{3+} ions (63 pm) can easily replace the Ti^{4+} ions (68 pm) from the TiO_2 crystal lattice. Further, Fe^{3+} ions prevent electron–hole (e^-–h^+) recombination, better charge separation and incorporation of oxygen vacancies in the crystal lattice and surface of the TiO_2 [28,29]. Additionally, the choice of Fe^{3+} ions as dopant has also been confirmed for some other semiconductor metal oxides. Hexamethylenediamine (HMDA) grafted and Fe nanoparticles incorporation into SnO_2 nanostructures exhibited strong surface affinity of 8.5 $\mu mol/m^2$ for H_2, as an important aspect of green energy storage applications at ambient temperature and pressure conditions [30]. ZnO powders doped with Fe nanoparticles through an in situ dispersion method showed improved conductivity and capacitance as compared to undoped ZnO nanoparticles [31]. HMDA grafted and Fe nanoparticles doped ZnO nanoparticles showed excellent conductivity which was attributed to the formation of effective proton-conductivity on the surface of the ZnO as well as proton transfer between Fe nanoparticles [32]. Fe-doped TiO_2 nanoparticles displayed superior photocatalytic degradation of methylene blue dye, phenol and toxic organic compounds as compared to undoped TiO_2 under UV and visible light illumination [33–35].

Herein, Fe-doped TiO_2 nanoparticles were prepared through a facile hydrothermal technique and subsequently analyzed using different characterization techniques for their various characteristics and to affirm the formation of the doped TiO_2 nanoparticles. Modified glassy carbon electrode (GCE) was fabricated by coating a thin film of Fe-doped TiO_2 nanoparticles onto it. As-fabricated hydrazine chemical sensors exhibited excellent hydrazine chemical sensing parameters.

2. Experimental Details

2.1. Synthesis of Fe-Doped TiO$_2$ Nanoparticles

For the preparation of Fe-doped TiO$_2$ nanoparticle, a facile hydrothermal method was adopted (2% doping of Fe ion was performed in the Fe-doped TiO$_2$ nanoparticle). Tetrabutyl titanate (Ti(BuO)$_4$) and ferric nitrate (Fe(NO$_3$)$_3$) were used as TiO$_2$ precursor and Fe dopant, respectively. Twenty milliliters of 0.5 M Ti(BuO)$_4$ solution was poured drop-wise to 5 M aqueous NaOH solution followed by the addition of 20 mL of 0.1 M Fe(NO$_3$)$_3$ solution dropwise. A Teflon lined stainless steel autoclave containing above solution was then heated at 200 °C for 10 h. The following reaction mechanisms have been proposed for the synthesis of Fe-doped TiO$_2$ nanoparticles (Equations (1)–(3)).

$$Ti(BuO)_4 + 4\,NaOH \rightarrow Ti(OH)_4 + 4\,BuONa \tag{1}$$

$$Fe(NO_3)_3 + 3\,NaOH \rightarrow Fe(OH)_3 + 3\,NaNO_3 \tag{2}$$

$$Ti(OH)_4 \xrightarrow{\text{Calcination}} TiO_2 + 2\,H_2O \tag{3}$$

During the hydrothermal process, tetrabutyl titanate was hydrolyzed to titanium hydroxide (Ti(OH)$_4$) and sodium butoxide. Ferric nitrate used as dopant was also hydrolyzed to its hydroxide, i.e., ferric hydroxide (Fe(OH)$_3$). BuONa and NaNO$_3$ were removed by washing with ethanol and deionized (DI) water. These washings also eliminated any unreacted Fe(NO$_3$)$_3$. Thoroughly washed Ti(OH)$_4$ and Fe(OH)$_3$ were dried for 6 h in an electric oven at 70 °C. During calcination, titanium hydroxide was converted into TiO$_2$ with Fe^{3+} ions occupying the lattice sites in the TiO$_2$ crystal lattice. The calcination was carried out in oxygen atmosphere at 450 °C for 2 h. The calculated synthesis yield of the synthesized Fe-doped TiO$_2$ nanoparticles were found to be 0.8 g.

2.2. Characterizations of Fe-Doped TiO$_2$ Nanoparticles

X-ray diffraction (XRD; PAN analytical Xpert Pro. Cambridge, UK) with Cu–Kα radiation was performed for the analysis of crystal phase and crystallite size. The optical characteristics were estimated by using UV–Vis Spectrophotometer (Perkin Elmer-Lamda 950, Waltham, MA, USA) within the wavelength range of 200–800 nm by dispersing and sonicating the Fe-doped TiO$_2$ in distilled water for 30 min. Scattering properties of the doped TiO$_2$ material were analyzed by Raman spectrum and examined using Raman spectrometer (Perkin Elmer-FTIR Spectrum-100, Waltham, MA, USA) from 100 to 900 cm^{-1}. FTIR spectrum of the Fe-doped TiO$_2$ was collected by FTIR spectrophotometer (Perkin Elmer-FTIR Spectrum-100, Waltham, MA, USA) through KBr pelletization from 500 to 4000 cm^{-1}. The photoelectronic properties of the Fe-doped TiO$_2$ nanoparticles were analyzed through photoluminescence spectral measurement.

2.3. Hydrazine Chemical Sensor Fabrication

Prior to electrode coating with Fe-doped TiO$_2$ nanoparticles, the GCE with surface area of 0.071 cm^2 (Bio-Logic SAS, Seyssinet-Pariset, France) was polished with a 1 µm polishing diamond. After this, the surface of the GCE was further polished with 0.05 µm alumina slurry and finally washed with distilled water. The modified GCEs with active materials (Fe-doped TiO$_2$) were fabricated as follows: The GCE surface was smoothly coated by the Fe-doped TiO$_2$ using ethyl acetate and conducting binder–butyl carbitol acetate followed by drying at 60 °C for 3 h. A three electrode electrochemical cell was connected to electrochemical workstation, Zahner Zennium, Germany, with a Pt wire as a counter electrode, and Fe-doped TiO$_2$ nanoparticle modified GCE as working electrode and an Ag/AgCl (saturated KCl) as a reference electrode were used during the electrochemical measurements. Different concentrations of hydrazine (0.2 µM–30 µM) were electrochemically tested. All the electroanalytical experiments were performed in 0.1 M phosphate buffer solution (PBS) of pH 7.4 at room temperature.

3. Results and Discussion

3.1. Characterizations and Properties of Fe-Doped TiO₂ Nanoparticles

Figure 1 depicts the XRD diffraction patterns of Fe-doped-TiO$_2$ nanoparticles. The XRD studies clearly demonstrated the presence of both anatase and rutile phases with anatase as the major phase. The presence of main diffraction peaks in the XRD pattern of Fe-doped-TiO$_2$ nanoparticles at 2θ = 25.28, 37.8, 48.07, 54.25 and 62.63, 68.9, 70.28 and 75.13 were consistent with (101), (103), (004), (200), (105), (211), (204), (116), (220) and (107) lattice planes of anatase phase (JCPDS No. 21-1272) [24].

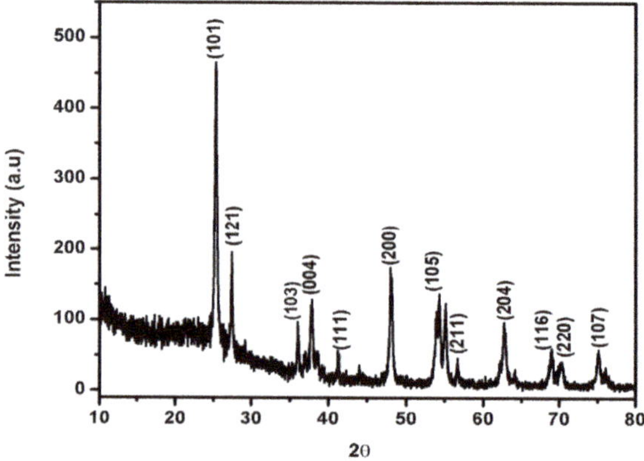

Figure 1. XRD pattern of Fe-doped TiO$_2$ nanoparticles.

The XRD peaks corresponding to rutile phase also emerged at 2θ = 27.50 and 41.54 diffraction angles corresponding to (121) and (111) diffraction planes (JCPDS No. 21-1276). The synthesized samples did not exhibit any diffraction peaks for Fe, which suggests that the Fe^{3+} content in the Fe-doped TiO$_2$ was below the detection limit, and due to almost similar ionic radii, the Fe^{3+} ions could substitute Ti^{4+} from some of the lattice sites of TiO$_2$ as discussed earlier. This further indicates that Fe^{3+} ions were successfully integrated into TiO$_2$ matrix homogeneously without the development of iron oxide on the TiO$_2$ surface. This homogeneous distribution of Fe^{3+} ions in TiO$_2$ matrix and low concentration, responsible for the absence of any Fe^{3+} peaks in the XRD patterns, have also been reported earlier in the literature [36,37]. The average crystallite size of the synthesized nanoparticles was estimated using the Scherrer formula (Equation (4)) and was found to be around 14 nm.

$$d = \frac{0.89\lambda}{\beta \cos\theta} \tag{4}$$

where λ = 1.542 Å, θ = Bragg angle of diffraction, β = full width at half maximum.

The detailed structural and morphological properties of Fe-doped TiO$_2$ were examined by FESEM analysis. The corresponding low and high magnification FESEM images are depicted in Figure 2a–d. A large number of spherical shaped and highly agglomerated Fe-doped TiO$_2$ particles with an average diameter of about 20 nm can be seen. In addition to spherical shapes, some Fe-doped TiO$_2$ with cubic, pentagonal, oval and other irregular geometries can also be seen from high magnification FESEM images (Figure 2c,d).

Figure 2. (**a**) and (**b**) Low magnification and (**c**) and (**d**) high magnification FESEM images for Fe-doped TiO$_2$ nanoparticles.

To observe the optical properties of the Fe-doped TiO$_2$ nanoparticles, a UV–Vis absorption spectroscopic study was performed. It can be examined from Figure 3a that Fe-doped TiO$_2$ nanoparticles exhibited a wide absorption peak below 400 nm, which is the typical absorption of TiO$_2$. This peak can be attributed to the electronic excitation from lower energy level to higher energy level in the anatase phase of the TiO$_2$. Furthermore, the change in color of the sample from white (pure TiO$_2$) to creamish yellow (Fe-doped TiO$_2$) depicted the increase in absorption towards visible light due to the incorporation of dopant metal [38].

Figure 3b illustrates the Raman spectra of Fe-doped TiO$_2$ nanoparticles. Five main bands—144.2, 195, 396.4, 513.2 and 634.8 cm^{-1}, corresponding to the six Raman active modes—were illustrated for the anatase phase of TiO$_2$ (3E$_g$, 2B$_{1g}$ and 1A$_{2g}$) [39]. The spectra indicated the crystalline nature of the synthesized nanoparticles. Furthermore, no additional peak related to the iron oxide was seen, which corroborates the results of XRD.

Figure 3c shows the FTIR spectra of Fe-doped TiO$_2$ nanoparticles. The spectrum reflects that doping had no effect on the bonding environment of the TiO$_2$ host nanoparticles. The broadband at 3433 cm^{-1} was assigned to the symmetric and asymmetric stretching vibrations of O–H bonds of the adsorbed water molecules during sample formation. An additional peak at 1628 cm^{-1} was attributed to the bending vibration related to the hydroxyl group of the adsorbed water. The band centered at ~520 cm^{-1} was due to metal–oxygen, i.e., Ti–O and Fe–O, bonds [40].

The photoelectronic properties of Fe-doped TiO$_2$ nanoparticles were studied from the photoluminescence spectrum (Figure 3d). The UV region peak at around 375 nm is likely related to the near-band-edge (NBE) excitonic emission. Interestingly, the energy corresponding to above NBE peak is close to the energy gap of 3.17 eV of anatase TiO$_2$ as reported earlier [40]. The NBE transition originated from the electrons–holes recombination. The incorporation of dopant Fe did not cause any significant alteration in the PL spectrum.

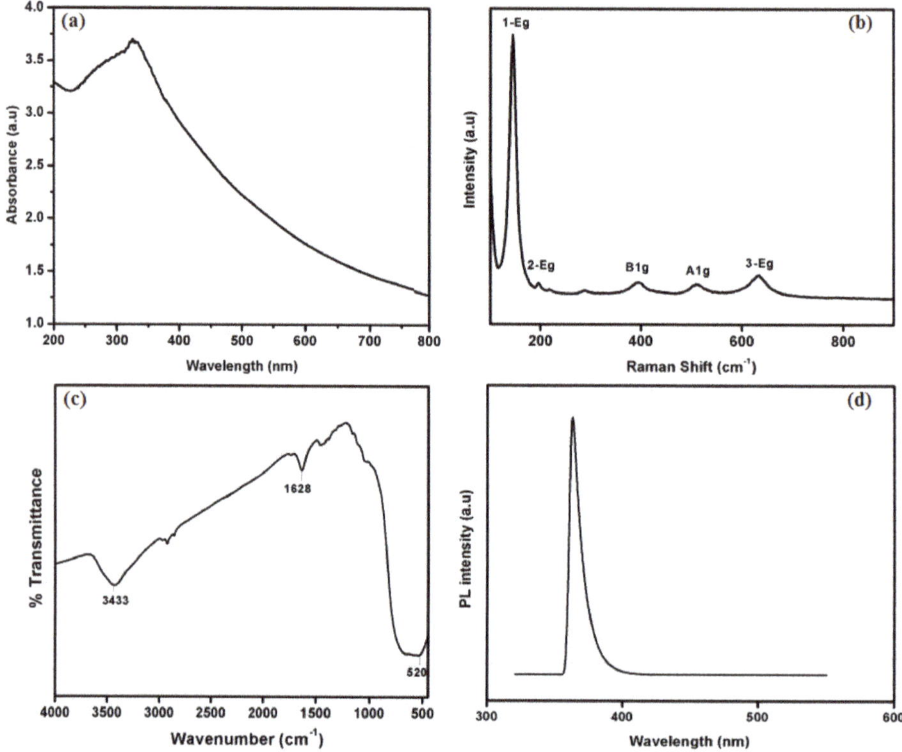

Figure 3. (**a**) UV–Vis absorption spectrum, (**b**) room temperature Raman spectrum, (**c**) FTIR and (**d**) PL spectrum of Fe-doped TiO$_2$ nanoparticles.

3.2. Electrochemical Sensing Properties of Hydrazine Using Fe-Doped TiO$_2$ Nanoparticles

The electrochemical sensing capability of Fe-doped TiO$_2$ nanostructure was examined using cyclic voltammetry. Figure 4a,b shows typical cyclic voltammogram (CV) of bare GCE, undoped TiO$_2$ modified GCE and Fe-doped TiO$_2$ modified GCE in absence and presence, respectively, of 0.5 mM hydrazine in 0.1 M PBS (pH 7.4) at 50 mVs^{-1} scan rate. As can be observed from Figure 4a, in blank PBS solution bare GCE, undoped TiO$_2$ modified GCE, and Fe-doped TiO$_2$ modified GCE did not generate any characteristic peak in the selected voltage range. However, with the addition of 0.5 mM hydrazine, no significant peak was observed by bare electrode, but a significant oxidation peak at 0.45 V vs. Ag/AgCl was detected in the case of both undoped TiO$_2$ modified GCE and Fe-doped TiO$_2$ modified electrode (Figure 4b), which clearly illustrated the efficient electrocatalytic activity of the modified GCE. The current response at Fe-doped TiO$_2$ modified electrode was much larger than the undoped TiO$_2$ modified GCE, ~137% larger, which indicates the enhanced electrocatalytic activity of coated GCE after the addition of Fe.

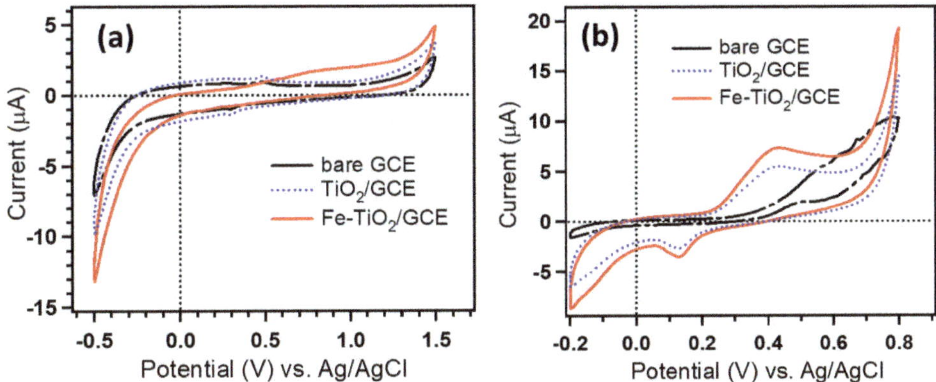

Figure 4. Cyclic voltammograms measured at 50 mVs^{-1} in 0.1 M phosphate buffer solution (PBS) (pH 7.4) (**a**) with absence of hydrazine and (**b**) in presence of 0.5 mM hydrazine using bare glassy carbon electrode (GCE), TiO$_2$ modified GCE and Fe-doped TiO$_2$ modified GCE.

3.3. Amperometric Studies

The amperometric (i–t) response was also carried out with the purpose to detect the hydrazine analyte using Fe-doped TiO$_2$ nanoparticle modified GCE. Amperometric studies were performed, and the constant potential was set at a value of 0.45 V with successive addition of hydrazine (0.2–30 µM) into a continuously stirred 0.1 M PBS (pH 7.4). As revealed in Figure 5a, Fe-doped TiO$_2$ modified GCE fabricated sensor illustrated a significant and steep rise in the current value after each successive addition of hydrazine. The measured current increased rapidly with a response time of ~20 s during the amperometric measurements. The corresponding calibration curve for amperometric hydrazine sensing showed a linear response in the concentration range 0.2 to 30 µM (Figure 5b). The correlation coefficient of the line R^2 was found to be 0.998, and the linear equation was y = 0.1019x + 0.0157. The linearity in the plot also confirmed that the hydrazine oxidation process was diffusion-controlled and indicated the fast electron transfer rate, which led to the sharper and well-defined peaks.

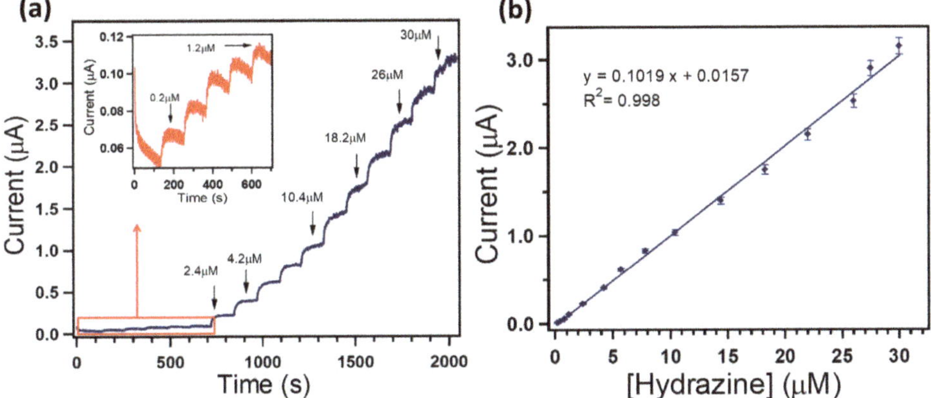

Figure 5. (**a**) Current–time (i–t) response of Fe-doped TiO$_2$ modified GCE for 0.2–30 µM hydrazine concentrations at a constant potential +0.45V vs. Ag/AgCl. Inset shows an enlarged part of the early stage addition (0.2–1.2 µM). (**b**) The corresponding current–concentration calibration graph.

It has been proposed that hydrazine in slightly basic medium (pH = 7.4) is oxidized onto the surface of the Fe-doped TiO_2 to release electrons (Equation (5)) [41,42].

$$N_2H_4 + 8\,HO^- \rightarrow 2\,NO + 6\,H_2O + 8\,e^- \tag{5}$$

The schematic proposed hydrazine sensing by Fe-doped TiO_2 is shown in Figure 6.

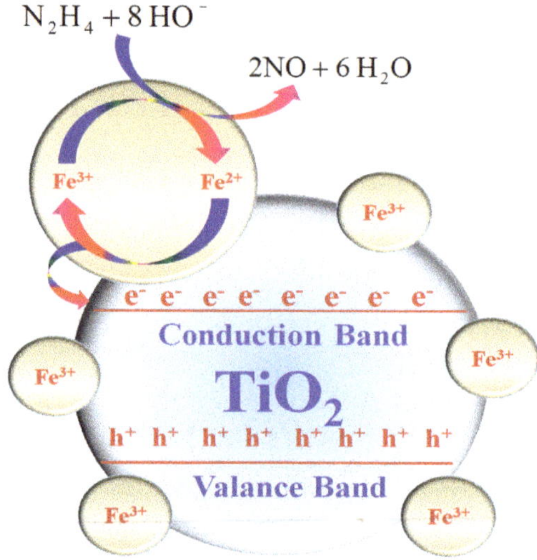

Figure 6. Proposed mechanism for the hydrazine sensing by Fe-doped TiO_2 nanoparticles.

The redox behavior of the Fe^{3+} ions facilitate the transfer of electrons released from the oxidation of the hydrazine to the conduction band of the TiO_2 or migrate the electrons to reduce the Ti^{4+} to Ti^{3+} ions. Li et al. [43] proposed that Fe^{3+} can act as a hole as well as an electron trap, which further enhances the electron transfer process. Furthermore, Zhu et al. [44] proposed the following reactions to depict the redox nature of Fe^{3+} ions ((Equations (6)–(9)).

$$Fe^{3+} + e^- \rightarrow Fe^{2+} \tag{6}$$

$$Fe^{3+} + h^+ \rightarrow Fe^{4+} \text{ (Hole trap)} \tag{7}$$

$$Fe^{3+} + Ti^{4+} \rightarrow Fe^{3+} + Ti^{3+} \text{ (Electron migration)} \tag{8}$$

$$Fe^{3+} + Ti^{3+} \rightarrow Fe^{2+} + Ti^{3+} \text{ (Recombination)} \tag{9}$$

Therefore, when hydrazine comes in contact with Fe-doped TiO_2, the electron density in the conduction band increases, which leads to the increase in electrical conductivity and, thus, increase in the current potential. The higher the concentration of the hydrazine, the greater is the electron density in the conduction band and, hence, the current potential is higher.

The sensitivity of the synthesized sensor was calculated from the slope of the calibration curve divided by the electrode area [45,46]. The limit of detection (LOD) of the fabricated hydrazine sensor was accordingly estimated via the following equation (Equation (10)):

$$LOD = 3.0^* \, \sigma_B/b \tag{10}$$

where σ_B is the standard deviation of the population of the blank signals (0.008 µA) and b is the slope of the regression line. It is possible to replace σ_B by the residual standard deviation of the regression,

$s_{y/x}$, also known as standard error of the regression [47]. The sensitivity and LOD were found to be 1.44 $\mu A\ \mu M^{-1}\ cm^{-2}$ and 0.236 μM, respectively. These obtained results demonstrated the potential of the Fe-doped TiO_2 nanostructure modified electrode as a suitable electrochemical sensor for sensitive and selective determination of hydrazine. The synthesized hydrazine electrochemical sensor was reliable and depicted fantastic reproducibility. It was ascertained that no significant decrease in sensitivity was observed when tested for more than three weeks while being stored at room temperature in a closed container.

In order to assess the analytical potential of the fabricated sensor for real sample analysis, the selectivity of Fe-doped TiO_2 nanostructures to hydrazine was evaluated. The selectivity test of the sensor was conducted to check the influence of some interfering electro-active chemical species on the sensing property of Fe-doped TiO_2 by measuring the i–t response. Figure 7 demonstrates the amperometric responses of Fe-doped TiO_2 modified GCE for the successive addition of different concentrations of hydrazine and 100 μM of ascorbic acid (AA), uric acid (UA), glucose, SO_4^{2-}, NO_3^{-}, Pb^{2+} and Ca^{2+} at a regular interval of 100 s in 0.1 M PBS at an applied potential value of +0.45 V. The negligible change in the observed current, even if the various co-existing interfering species were present, undoubtedly revealed the excellent selectivity of the fabricated sensor.

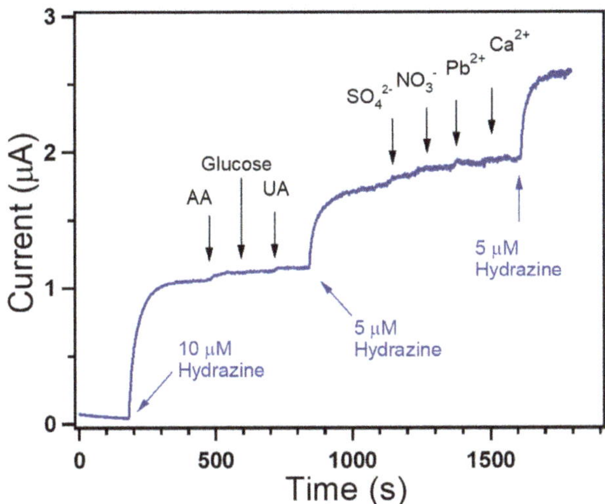

Figure 7. Amperometric (i–t) measurement showing the interference behavior of the Fe-doped TiO_2 coated GCE upon the successive injections of 10 or 5 μM hydrazine and each 100 μM of AA, glucose, UA, SO_4^{2-}, NO_3^{-}, Pb^{2+} and Ca^{2+} into a continuously stirred 0.1 M PBS (pH 7.4) solution operating at +0.45 V vs. Ag/AgCl.

The sensor parameters of Fe-doped TiO_2 nanoparticle modified GCE and other recently reported sensors are compared in Table 1. Detailed comparison shows that the fabricated sensor had excellent electrocatalytic performance for selective detection and sensing of hydrazine.

Table 1. Comparison of the sensor parameters of the Fe-doped TiO_2 coated GCE sensor with some recently reported hydrazine electrochemical sensor materials.

Sensor Electrode	Sensitivity	LOD	Ref.
Leaf shape CuO/OMC/GCE	$0.00487\ \mu A.\mu M^{-1}.cm^{-2}$	$0.887\ \mu M$	[4]
$NiCo_2S_4$ sphere/GCE	$179.1\mu A.mM^{-1}.cm^{-2}$	$0.6\mu M$	[5]
WO_3 NPs/Au electrode	$0.185\ \mu A.\mu M^{-1}.cm^{-2}$	$144.73\ \mu M$	[32]
Polythiophene/ZnO/GCE	$1.22\ \mu A.\mu M^{-1}.cm^{-2}$	$0.207\mu M$	[48]
Ag@Fe_3O_4 nanosphere/GCE	$270.0\mu A.mM^{-1}.cm^{-2}$	$0.06\mu M$	[49]
CdO/CNT nanocomposites/GCE	$25.79\mu A.\mu M^{-1}.cm^{-2}$	$4.0\ pM$	[50]
Pd/Co-NCNTs	$343.9\ \mu A.mM^{-1}.cm^{-2}$	$0.007\ \mu M$	[51]
Pd (CNT-Pd)/GCE	$0.3\ \mu A.mM^{-1}.cm^{-2}$	$8.0\ \mu M$	[52]
Nanoporous gold/ITO	$0.161\ \mu A.\mu M^{-1}.cm^{-2}$	$0.0043\ \mu M$	[53]
α-Fe_2O_3/polyaniline nanocomposite	$1.93\ mA.\mu M^{-1}.cm^{-2}$	$0.153\ \mu M$	[54]
Chrysanthemum-like Co_3O_4/GCE	$107.9\ \mu A.mM^{-1}.cm^{-2}$	$3.7\ \mu M$	[55]
AuNPs/CNTs-rGO/GCE	$9.73\ \mu A.\mu M^{-1}.cm^{-2}$	$0.065\ \mu M$	[56]
Fe-doped TiO_2 NPs/GCE	$1.44\ \mu A.\mu M^{-1}.cm^{-2}$	$0.236\ \mu M$	This work

4. Conclusions

In summary, highly crystalline Fe-doped TiO_2 nanoparticles were synthesized through hydrothermal synthesis and subsequently characterized by different characterization techniques. Finally, Fe-doped TiO_2 nanoparticles were applied as an efficient electron mediator for the fabrication of hydrazine chemical sensor using GCE. As-fabricated modified GCE showed sensitivity, LOD and LDR of $1.44\ \mu A\ \mu M^{-1}\ cm^{-2}$, $0.236\ \mu M$ and 0.2–$30\ \mu M$, respectively, through an amperometric sensing approach. It was proposed that redox behavior of the Fe^{3+} ions facilitates the transfer of electrons released from the oxidation of the hydrazine to the conduction band of the TiO_2, which leads to the increase in electrical conductivity. The negligible change in the observed current in the presence of various co-existing interfering chemical species further confirms the excellent selectivity of the fabricated sensor towards hydrazine.

Author Contributions: A.U., F.A.H., A.A.I., T.A., R.K., M.S.A.-A. and S.B. conceived of the presented idea and developed and performed the experiments. All of the authors discussed the results and contributed to the final manuscript. All authors have read and agreed to the published version of the manuscript.

Funding: This research was funded by the Ministry of Education, Kingdom of Saudi Arabia through a grant (PCSED-013-18) under the Promising Centre for Sensors and Electronic Devices (PCSED) at Najran University, Kingdom of Saudi Arabia.

Acknowledgments: Authors would like to acknowledge the support of the Ministry of Education, Kingdom of Saudi Arabia for this research through a grant (PCSED-013-18) under the Promising Centre for Sensors and Electronic Devices (PCSED) at Najran University, Kingdom of Saudi Arabia.

Conflicts of Interest: The authors declare no conflict of interest.

References

1. Wahab, R.; Ahmad, N.; Alam, M.; Ahmed, J. Nanorods of ZnO: An effective hydrazine sensor and their chemical properties. *Vacuum* **2019**, *165*, 290–296. [CrossRef]
2. Guo, S.-H.; Guo, Z.-Q.; Wang, C.-Y.; Shen, Y.; Zhu, W.-H. An ultrasensitive fluorescent probe for hydrazine detection and its application in water samples and living cells. *Tetrahedron* **2019**, *75*, 2642–2646. [CrossRef]
3. Nemakal, M.; Aralekallu, S.; Mohammed, I.; Swamy, S.; Sannegowda, L.K. Electropolymerized octabenzimidazole phthalocyanine as an amperometric sensor for hydrazine. *J. Electroanal. Chem.* **2019**, *839*, 238–246. [CrossRef]
4. Wang, L.; Meng, T.; Jia, H.; Feng, Y.; Gong, T.; Wang, H.; Zhang, Y. Electrochemical study of hydrazine oxidation by leaf-shaped copper oxide loaded on highly ordered mesoporous carbon composite. *J. Colloid Interface Sci.* **2019**, *549*, 98–104. [CrossRef]

5. Duan, C.; Dong, Y.; Sheng, Q.; Zheng, J. A high-performance non-enzymatic electrochemical hydrazine sensor based on NiCo2S4 porous sphere. *Talanta* **2019**, *198*, 23–29. [CrossRef]

6. George, M.; Nagaraja, K.S.; Balasubramanian, N. Spectrophotometric determination of hydrazine. *Talanta* **2008**, *75*, 27–31. [CrossRef]

7. Rios, A.; Silva, M.; Valcarcel, M. Fluorimetric determination of ammonia, hydrazine and hydroxylamine and their mixtures by differential kinetic methods. *Fresenius' Zeitschrift Für Anal. Chem.* **1985**, *320*, 762–768. [CrossRef]

8. Collins, G.E.; Rose-Pehrsson, S.L. Sensitive, fluorescent detection of hydrazine via derivatization with 2,3-naphthalene dicarboxaldehyde. *Anal. Chim. Acta* **1993**, *284*, 207–215. [CrossRef]

9. Collins, G.E.; Latturner, S.; Rose-Pehrsson, S.L. Chemiluminescence detection of hydrazine vapor. *Talanta* **1995**, *42*, 543–551. [CrossRef]

10. Bravo, P.; Isaacs, F.; Ramírez, G.; Azócar, I.; Trollund, E.; Aguirre, M.J. A potentiometric hydrazine sensor: Para-Ni-tetraaminophenylporphyrin/Co-cobaltite/SNO2:F modified electrode. *J. Coord. Chem.* **2007**, *60*, 2499–2507. [CrossRef]

11. Ganesh, S.; Khan, F.; Ahmed, M.K.; Pandey, S.K. Potentiometric determination of free acidity in presence of hydrolysable ions and a sequential determination of hydrazine. *Talanta* **2011**, *85*, 958–963. [CrossRef] [PubMed]

12. Kannan, P.K.; Moshkalev, S.A.; Rout, C.S. Electrochemical sensing of hydrazine using multilayer graphene nanobelts. *RSC Adv.* **2016**, *6*, 11329–11334. [CrossRef]

13. Rostami, S.; Azizi, S.N.; Ghasemi, S. Simultaneous electrochemical determination of hydrazine and hydroxylamine by CuO doped in ZSM-5 nanoparticles as a new amperometric sensor. *New J. Chem.* **2017**, *41*, 13712–13723. [CrossRef]

14. Jena, B.K.; Raj, C.R. Ultrasensitive nanostructured platform for the electrochemical sensing of hydrazine. *J. Phys. Chem. C* **2007**, *111*, 6228–6232. [CrossRef]

15. Deng, J.; Deng, S.; Liu, Y. Highly sensitive electrochemical sensing platform for hydrazine detection. *Int. J. Electrochem. Sci.* **2018**, *13*, 3566–3574. [CrossRef]

16. Gu, X.; Li, X.; Wu, S.; Shi, J.; Jiang, G.; Jiang, G.; Tian, S. A sensitive hydrazine hydrate sensor based on a mercaptomethyl-terminated trinuclear Ni(ii) complex modified gold electrode. *RSC Adv.* **2016**, *6*, 8070–8078. [CrossRef]

17. Wang, J.; Wang, H.; Yang, S.; Tian, H.; Liu, Y.; Hao, Y.; Zhang, J.; Sun, B. A Fluorescent Probe for Sensitive Detection of Hydrazine and Its Application in Red Wine and Water. *Anal. Sci.* **2018**, *34*, 329–333. [CrossRef]

18. Yang, Y.J. Electrodeposition of ZnO Nanofibers for Sensitive Determination of Hydrazine and Nitrite. *Sens. Lett.* **2017**, *14*, 1187–1192. [CrossRef]

19. Ma, Y.; Li, H.; Wang, R.; Wang, H.; Lv, W.; Ji, S. Ultrathin willow-like CuO nanoflakes as an efficient catalyst for electro-oxidation of hydrazine. *J. Power Sources* **2015**, *289*, 22–25. [CrossRef]

20. Wang, G.; Gu, A.; Wang, W.; Wei, Y.; Wu, J.; Wang, G.; Zhang, X.; Fang, B. Copper oxide nanoarray based on the substrate of Cu applied for the chemical sensor of hydrazine detection. *Electrochem. Commun.* **2009**, *11*, 631–634. [CrossRef]

21. Wang, G.; Zhang, C.; He, X.; Li, Z.; Zhang, X.; Wang, L.; Fang, B. Detection of hydrazine based on Nano-Au deposited on Porous-TiO2 film. *Electrochim. Acta* **2010**, *55*, 7204–7210. [CrossRef]

22. Yi, Q.; Niu, F.; Yu, W. Pd-modified TiO2 electrode for electrochemical oxidation of hydrazine, formaldehyde and glucose. *Thin Solid Films* **2011**, *519*, 3155–3161. [CrossRef]

23. Wu, J.; Zhou, T.; Wang, Q.; Umar, A. Morphology and chemical composition dependent synthesis and electrochemical properties of MnO2-based nanostructures for efficient hydrazine detection. *Sens. Actuators B Chem.* **2016**, *224*, 878–884. [CrossRef]

24. Shirsath, S.R.; Pinjari, D.V.; Gogate, P.R.; Sonawane, S.H.; Pandit, A.B. Ultrasound assisted synthesis of doped TiO 2 nano-particles: Characterization and comparison of effectiveness for photocatalytic oxidation of dyestuff effluent. *Ultrason. Sonochem.* **2013**, *20*, 277–286. [CrossRef]

25. Khodari, M.; Mersal, G.A.M.; Rabie, E.M.; Assaf, H.F. Electrochemical sensor based on carbon paste electrode modified by TiO2 nano-particles for the voltammetric determination of resorcinol. *Int. J. Electrochem. Sci.* **2018**, *13*, 3460–3474. [CrossRef]

26. Guo, Q.; Zhang, M.; Li, X.; Li, X.; Li, H.; Lu, Y.; Song, X.; Wang, L. A novel CuO/TiO 2 hollow nanofiber film for non-enzymatic glucose sensing. *RSC Adv.* **2016**, *6*, 99969–99976. [CrossRef]

27. Guo, Q.; Liu, L.; Zhang, M.; Hou, H.; Song, Y.; Wang, H.; Zhong, B.; Wang, L. Hierarchically mesostructured porous TiO2 hollow nanofibers for high performance glucose biosensing. *Biosens. Bioelectron.* **2017**, *92*, 654–660. [CrossRef]

28. Spriano, S.; Yamaguchi, S.; Baino, F.; Ferraris, S. A critical review of multifunctional titanium surfaces: New frontiers for improving osseointegration and host response, avoiding bacteria contamination. *Acta Biomater.* **2018**, *79*, 1–22. [CrossRef]

29. Zhang, Y.; Cheng, K.; Lv, F.; Huang, H.; Fei, B.; He, Y.; Ye, Z.; Shen, B. Photocatalytic treatment of 2,4,6-trinitotoluene in red water by multi-doped TiO$_2$ with enhanced visible light photocatalytic activity. *Colloids Surf. A Physicochem. Eng. Asp.* **2014**, *452*, 103–108. [CrossRef]

30. Bouazizi, N.; Bargougui, R.; Boudharaa, T.; Khelil, M.; Benghnia, A.; Labiadh, L.; Slama, R.B.; Chaouachi, B.; Ammar, S.; Azzouz, A. Synthesis and characterization of SnO$_2$-HMD-Fe materials with improved electric properties and affinity towards hydrogen. *Ceram. Int.* **2016**, *42*, 9413–9418. [CrossRef]

31. Bouazizi, N.; Ajala, F.; Khelil, M.; Lachheb, H.; Khirouni, K.; Houas, A.; Azzouz, A. Zinc oxide incorporating iron nanoparticles with improved conductance and capacitance properties. *J. Mater. Sci. Mater. Electron.* **2016**, *27*, 11168–11175. [CrossRef]

32. Bouazizi, N.; Ajala, F.; Bettaibi, A.; Khelil, M.; Benghnia, A.; Bargougui, R.; Louhichi, S.; Labiadh, L.; Slama, R.B.; Chaouachi, B.; et al. Metal-organo-zinc oxide materials: Investigation on the structural, optical and electrical properties. *J. Alloys Compd.* **2016**, *656*, 146–153. [CrossRef]

33. Khan, M.A.M.; Siwach, R.; Kumar, S.; Alhazaa, A.N. Role of Fe doping in tuning photocatalytic and photoelectrochemical properties of TiO$_2$ for photodegradation of methylene blue. *Opt. Laser Technol.* **2019**, *118*, 170–178. [CrossRef]

34. Moradi, V.; Ahmed, F.; Jun, M.B.G.; Blackburn, A.; Herring, R.A. Acid-treated Fe-doped TiO$_2$ as a high performance photocatalyst used for degradation of phenol under visible light irradiation. *J. Environ. Sci. China* **2019**, *83*, 183–194. [CrossRef]

35. Sood, S.; Umar, A.; Mehta, S.K.; Kansal, S.K. Highly effective Fe-doped TiO$_2$ nanoparticles photocatalysts for visible-light driven photocatalytic degradation of toxic organic compounds. *J. Colloid Interface Sci.* **2015**, *450*, 213–223. [CrossRef]

36. Ismael, M. Enhanced photocatalytic hydrogen production and degradation of organic pollutants from Fe (III) doped TiO$_2$ nanoparticles. *J. Environ. Chem. Eng.* **2020**, *8*, 103676. [CrossRef]

37. Khan, M.A.; Woo, S.I.; Yang, O.B. Hydrothermally stabilized Fe(III) doped titania active under visible light for water splitting reaction. *Int. J. Hydrogen Energy* **2008**, *33*, 5345–5351. [CrossRef]

38. Niu, Y.; Xing, M.; Zhang, J.; Tian, B. Visible light activated sulfur and iron co-doped TiO$_2$ photocatalyst for the photocatalytic degradation of phenol. *Catal. Today* **2013**, *201*, 159–166. [CrossRef]

39. Ali, I.; Kim, S.-R.; Park, K.; Kim, J.-O. One-step electrochemical synthesis of graphene oxide-TiO$_2$ nanotubes for improved visible light activity. *Opt. Mater. Express* **2017**, *7*, 1535. [CrossRef]

40. Prajapati, B.; Kumar, S.; Kumar, M.; Chatterjee, S.; Ghosh, A.K. Investigation of the physical properties of Fe:TiO$_2$ -diluted magnetic semiconductor nanoparticles. *J. Mater. Chem. C* **2017**, *5*, 4257–4267. [CrossRef]

41. Dong, B.; He, B.L.; Huang, J.; Gao, G.Y.; Yang, Z.; Li, H.L. High dispersion and electrocatalytic activity of Pd/titanium dioxide nanotubes catalysts for hydrazine oxidation. *J. Power Sources* **2008**, *175*, 266–271. [CrossRef]

42. Shukla, S.; Chaudhary, S.; Umar, A.; Chaudhary, G.R.; Mehta, S.K. Tungsten oxide (WO$_3$) nanoparticles as scaffold for the fabrication of hydrazine chemical sensor. *Sens. Actuators B Chem.* **2014**, *196*, 231–237. [CrossRef]

43. Li, Z.; Shen, W.; He, W.; Zu, X. Effect of Fe-doped TiO$_2$ nanoparticle derived from modified hydrothermal process on the photocatalytic degradation performance on methylene blue. *J. Hazard. Mater.* **2008**, *155*, 590–594. [CrossRef]

44. Zhu, J.; Zheng, W.; He, B.; Zhang, J.; Anpo, M. Characterization of Fe-TiO$_2$ photocatalysts synthesized by hydrothermal method and their photocatalytic reactivity for photodegradation of XRG dye diluted in water. *J. Mol. Catal. A Chem.* **2004**, *216*, 35–43. [CrossRef]

45. Dong, B.; Chai, Y.M.; Liu, Y.Q.; Liu, C.G. Facile Synthesis and High Activity of Novel Ag/TiO$_2$-NTs Composites for Hydrazine Oxidation. *Adv. Mater. Res.* **2011**, *197*, 1073–1078. [CrossRef]

46. Dong, B.; He, B.L.; Chai, Y.M.; Liu, C.G. Novel Pt nanoclusters/titanium dioxide nanotubes composites for hydrazine oxidation. *Mater. Chem. Phys.* **2010**, *120*, 404–408. [CrossRef]

47. Miller, J.N.; Miller, J.N. *Statistics and Chemometrics for Analytical Chemistry*, 5th ed.; Pearson Prentice Hall: Edinburgh, Scotland, 2005.
48. Faisal, M.; Harraz, F.A.; Al-Salami, A.E.; Al-Sayari, S.A.; Al-Hajry, A.; Al-Assiri, M.S. Polythiophene/ZnO nanocomposite-modified glassy carbon electrode as efficient electrochemical hydrazine sensor. *Mater. Chem. Phys.* **2018**, *214*, 126–134. [CrossRef]
49. Dong, Y.; Yang, Z.; Sheng, Q.; Zheng, J. Solvothermal synthesis of Ag@Fe$_3$O$_4$ nanosphere and its application as hydrazine sensor. *Colloids Surf. A Physicochem. Eng. Asp.* **2018**, *538*, 371–377. [CrossRef]
50. Rahman, M.M.; Alam, M.M.; Alamry, K.A. Sensitive and selective m-tolyl hydrazine chemical sensor development based on CdO nanomaterial decorated multi-walled carbon nanotubes. *J. Ind. Eng. Chem.* **2019**, *77*, 309–316. [CrossRef]
51. Zhang, Y.; Huang, B.; Ye, J.; Ye, J. A sensitive and selective amperometric hydrazine sensor based on palladium nanoparticles loaded on cobalt-wrapped nitrogen-doped carbon nanotubes. *J. Electroanal. Chem.* **2017**, *801*, 215–223. [CrossRef]
52. Gioia, D.; Casella, I.G. Pulsed electrodeposition of palladium nano-particles on coated multi-walled carbon nanotubes/nafion composite substrates: Electrocatalytic oxidation of hydrazine and propranolol in acid conditions. *Sens. Actuators B Chem.* **2016**, *237*, 400–407. [CrossRef]
53. Tang, Y.-Y.; Kao, C.-L.; Chen, P.-Y. Electrochemical detection of hydrazine using a highly sensitive nanoporous gold electrode. *Anal. Chim. Acta* **2012**, *711*, 32–39. [CrossRef] [PubMed]
54. Harraz, F.A.; Ismail, A.A.; Al-Sayari, S.A.; Al-Hajry, A.; Al-Assiri, M.S. Highly sensitive amperometric hydrazine sensor based on novel α-Fe$_2$O$_3$/crosslinked polyaniline nanocomposite modified glassy carbon electrode. *Sens. Actuators B Chem.* **2016**, *234*, 573–582. [CrossRef]
55. Zhou, T.; Lu, P.; Zhang, Z.; Wang, Q.; Umar, A. Perforated Co$_3$O$_4$ nanoneedles assembled in chrysanthemum-like Co3O4 structures for ultra-high sensitive hydrazine chemical sensor. *Sens. Actuators B Chem.* **2016**, *235*, 457–465. [CrossRef]
56. Zhao, Z.; Sun, Y.; Li, P.; Zhang, W.; Lian, K.; Hu, J.; Chen, Y. Preparation and characterization of AuNPs/CNTs-ErGO electrochemical sensors for highly sensitive detection of hydrazine. *Talanta* **2016**, *158*, 283–291. [CrossRef]

Article

ZnO Nanocrystal-Based Chloroform Detection: Density Functional Theory (DFT) Study

H. Y. Ammar [1,2,3,*], H. M. Badran [1,2,3], Ahmad Umar [1,4,*], H. Fouad [5,6,*] and Othman Y. Alothman [7]

[1] Promising Centre for Sensors and Electronic Devices (PCSED), Najran University, Najran 11001, Saudi Arabia; drhebabadran@gmail.com
[2] Department of Physics, Faculty of Science and Arts, Najran University, Najran 11001, Saudi Arabia
[3] Physics Department, Faculty of Education, Ain Shams University, Cairo 11566, Egypt
[4] Department of Chemistry, Faculty of Science and Arts, Najran University, Najran 11001, Saudi Arabia
[5] Department of Applied Medical Science, Community College, King Saud University, Riyadh 11437, Saudi Arabia
[6] Biomedical Engineering Department, Faculty of Engineering, Helwan University, P.O. Box, Helwan 11792, Egypt
[7] Chemical Engineering Department, King Saud University, P. O. Box 800, Riyadh 11421, Saudi Arabia; othman@ksu.edu.sa
[*] Correspondence: hyammar@hotmail.com (H.Y.A.); umahmad@nu.edu.sa or ahmadumar786@gmail.com (A.U.); menhfef@ksu.edu.sa (H.F.)

Received: 5 October 2019; Accepted: 14 November 2019; Published: 19 November 2019

Abstract: We investigated the detection of chloroform ($CHCl_3$) using ZnO nanoclusters via density functional theory calculations. The effects of various concentrations of $CHCl_3$, as well as the deposition of O atoms, on the adsorption over ZnO nanoclusters were analyzed via geometric optimizations. The calculated difference between the highest occupied molecular orbital and the lowest unoccupied molecular orbital for ZnO was 4.02 eV. The most stable adsorption characteristics were investigated with respect to the adsorption energy, frontier orbitals, elemental positions, and charge transfer. The results revealed that ZnO nanoclusters with a specific geometry and composition are promising candidates for chloroform-sensing applications.

Keywords: nanocrystal; ZnO; density of states; optical and electrical properties

---

1. Introduction

The rapid development of various important industries, such as automobiles, pharmaceuticals, textiles, food, and agriculture, has substantially contributed to environmental pollution [1]. The release of various toxic and harmful gases and chemicals from such industries has significantly disturbed the ecosystem, and poses a great threat not only to humans, but to all living beings [2,3]. Among the various toxic gases, chloroform, which is also known as tri-chloromethane or methyl-tri-chloride, is considered to be one of the most toxic gases, and evaporates quickly when exposed to air [4]. It is widely used by chemical companies and in paper mills. Chloroform lasts for a long time in the environment, and its breakdown products, such as phosgene and hydrogen chloride, are as toxic or even more toxic [5]. The exposure of humans to chloroform severely affects the central nervous system, kidneys, liver, etc. Long-term exposure may result in vomiting, nausea, dizziness, convulsions, depression, respiratory failure, coma, and even sudden death [6,7]. It is important to efficiently detect the release of chloroform because of its serious health hazards. Thus, various methods have been reported for detecting chloroform, which involve optical sensors, colorimetric sensors, fluorescent sensors, electrochemical sensors, resistive gas sensors, luminescent sensors, photo-responsive sensors, etc. [8–13].

Among the various sensing techniques, gas sensors have attracted considerable attention because of their facile manufacturing process, high sensitivity, and low detection limit [14–17]. The literature reveals that metal-oxide materials are the most widely used scaffold to fabricate gas sensors [15–20]. In particular, metal-oxide materials are widely utilized to fabricate sensors for toxic and explosive gases [20–24]. It has been observed that the nanocrystal interfaces can significantly influence the optical and electrical properties and charge-trapping phenomena [22–25]. Zinc oxide (ZnO) is one of the most important and functional materials because of its various significant properties, including its wide bandgap; high exciton binding energy, piezoelectricity, and pyroelectricity; high conductivity and electron mobility; good stability in chemical and thermal environments; and biocompatibility [26–28]. Therefore, to improve the gas-sensing performance of ZnO-based gas sensors, various approaches have been employed, such as doping, surface modification, and the fabrication of composites [26,29]. Although ZnO materials are used for various gas-sensing applications [30–34], there are few reports available on the use of ZnO materials for chloroform sensing. Ghenaatian et al. [35] investigated the $Zn_{12}O_{12}$ nanocage as a promising adsorbent and detector for CS_2. Baie et al. examined the $Zn_{12}O_{12}$ fullerene-like cage as a potential sensor for SO_2 detection [36]. Ammar [37] reported that the $Zn_{12}O_{12}$ nanocage is a potential sorbent and detector for formaldehyde molecules. Nanocrystalline ZnO thin-film gas sensors were investigated by Mayya et al. [38] for the detection of hydrochloric acid, ethanolamine, and chloroform. Additionally, it is important to examine various geometries and other electronic parameters in order to obtain the optimal sensing material based on ZnO.

In this study, we investigated the detection of chloroform ($CHCl_3$) using ZnO nanoclusters via density functional theory (DFT) calculations implemented in a Gaussian 09 program. The effect of various concentrations of $CHCl_3$, as well as the deposition of O atoms, on the adsorption over ZnO nanoclusters was analyzed via geometric optimizations. To fully exploit the ZnO nanocrystals, various calculations related to the gas-sensing properties were performed.

2. Methods and Computational Details

A quantum cluster consisting of 24 atoms ($Zn_{12}O_{12}$) was selected to study the interaction between the ZnO nanocage and the $CHCl_3$ molecule. DFT calculations were performed with the Gaussian 09 suite of programs [39]. The calculations were conducted using Becke's three-parameter B3 with the Lee, Yang, and Parr (LYP) correlation functional [40]. This B3LYP hybrid functional contains the exchange–correlation functional, and is based on the exact form of the Vosko–Wilk–Nusair correlation potential [41]. Originally, the functional B included the Slater exchange along with corrections involving the gradient of density [42]. The correlation functional LYP was that of Lee, Yang, and Parr, which includes both local and nonlocal terms [43,44]. For the ZnO nanocage, the standard LANL2DZ basis set [37,45] was used. For the $CHCl_3$ and the deposited O atoms, a 6-31G (d, p) basis set was used. The adsorption energy (E_{ads}) of the $CHCl_3$ molecule on the surface of the $Zn_{12}O_{12}$ nanocage is defined as follows:

$$E_{ads} = \left[E_{(CHCl_3)_n/ZnO} - \left(nE_{CHCl_3} + E_{ZnO} \right) \right]/n, \tag{1}$$

where E_{CHCl_3}, E_{ZnO} and $E_{(CHCl_3)_n/ZnO}$ represent the energies of a single $CHCl_3$ molecule, the pristine $Zn_{12}O_{12}$ nanocage, and the $(CHCl_3)_n/Zn_{12}O_{12}$ complex, respectively.

The adsorption energy (E_{ads}) of an O atom on the surface of the $Zn_{12}O_{12}$ nanocage is defined as follows:

$$E_{ads} = \left[E_{O_n/ZnO} - \left(nE_O + E_{ZnO} \right) \right]/n, \tag{2}$$

where E_O and $E_{O_n/ZnO}$ represent the energies of a single O atom and the $O_n/Zn_{12}O_{12}$ complex, respectively.

The adsorption energy (E_i) of a $CHCl_3$ molecule on the deposited O on the $Zn_{12}O_{12}$ nanocage is defined as follows:

$$E_i = \left[E_{(CHCl_3)_n/O_n/ZnO} - \left(nE_{CHCl_3} + nE_O + E_{ZnO} \right) \right]/n, \tag{3}$$

where $E_{(CHCl_3)_n/O_n/ZnO}$ represents the energy of the $(CHCl_3)_n/O_n/ZnO$ complex.

The positive and negative values of E_i indicate the endothermic and exothermic processes, respectively. The binding energy (E_b) between the X and Y fragments of the XY complex is defined as follows:

$$E_b = E_{XY} - (E_X + E_Y),\qquad(4)$$

where E_{XY} represents the total energy of the optimized molecule, and E_X and E_Y represent the energies of the two fragments X and Y, respectively, having the same geometric structure as in the XY complex. The GaussSum 2.2.5 program was used to calculate the densities of states (DOSs) for the $Zn_{12}O_{12}$ nanocage, $CHCl_3$, and other complex systems [46]. Full natural bond orbital (NBO; NBO version 3.1) analyses were used to estimate the charge distributions for the $Zn_{12}O_{12}$ nanocages, $CHCl_3$, and other complex systems [47].

3. Results and Discussion

3.1. Geometric Optimization

The geometric optimization of a pristine $Zn_{12}O_{12}$ nanocage was performed. $Zn_{12}O_{12}$ is composed of eight $(ZnO)_3$ and six $(ZnO)_2$ rings, forming a cluster in which all of the Zn and O vertices are equivalent [48], as shown in Figure 1. The examined structural properties of the $Zn_{12}O_{12}$ nanocage agreed well with previous studies [37,45,49]. For example, the bond lengths $R_{Zn\text{-}O}$ of 1.91 and 1.98 Å were close to the previously reported values of 1.91 and 1.98 Å, respectively [37,45], and 1.89 and 1.97 Å, respectively [47]. The calculated highest occupied molecular orbital (HOMO)–lowest unoccupied molecular orbital (LUMO) energy gap for ZnO was found to be 4.02 eV, which agrees well with a previous work [45].

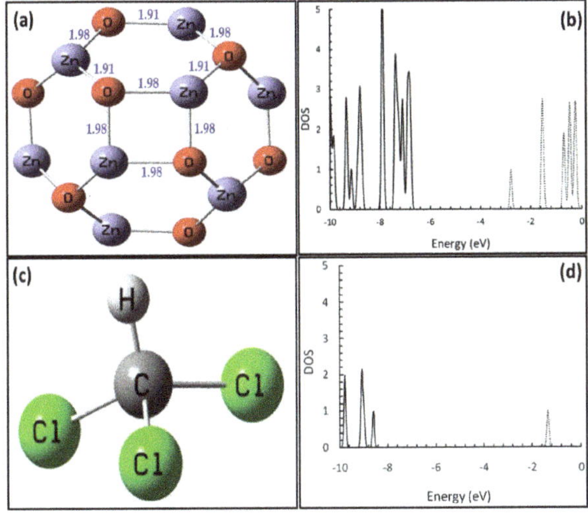

Figure 1. Optimized structures and densities of states (DOSs) of the ZnO nanocage ($Zn_{12}O_{12}$) and $CHCl_3$ used in the calculations. (**a,b**) Optimized structure and DOS of the ZnO nanocage; (**c,d**) optimized structure and DOS of $CHCl_3$. The distances are in Å, and the DOS is in arbitrary units. The solid and dashed lines represent the occupied and virtual states, respectively.

The DOS of the $Zn_{12}O_{12}$ nanocage was calculated, as shown in Figure 1. A geometric optimization was performed for the chloroform molecule ($CHCl_3$). It is a tetrahedral molecule, as shown in Figure 1. The calculated structural properties of $CHCl_3$ indicated that bond lengths $R_{C\text{-}H}$ and $R_{C\text{-}Cl}$ were 1.09 and 1.79 Å, respectively, and angles $A_{H\text{-}C\text{-}Cl}$ and $A_{Cl\text{-}C\text{-}Cl}$ were 107.5° and 11.4°, respectively. The energy gap (E_g) between the HOMO and LUMO was calculated to be 7.27 eV. The DOS for $CHCl_3$ was calculated, and is presented in Figure 1.

3.2. $CHCl_3$ Interaction with the $Zn_{12}O_{12}$ Nanocage

The geometric optimizations for four probable orientations of $CHCl_3$ on the surface of the $Zn_{12}O_{12}$ nanocage were investigated. Figure 2 shows the four orientations where the $CHCl_3$ molecule may interact via its H head or Cl head, and may be absorbed over the O site or Zn site of the $Zn_{12}O_{12}$ nanocage. The adsorption energy was calculated using Equation (1). The electronic properties of the $CHCl_3$ adsorption modes are presented in Table 1. For the first adsorption mode (a), the $CHCl_3$ molecule was weakly chemically adsorbed, and for the other modes (b, c, and d), the $CHCl_3$ molecule was physically adsorbed. The boundary value between the physical and chemical adsorption was considered to be 0.21 eV [50,51]. In mode (a), owing to the chemical interaction, the Fermi level (E_{FL}) for the cluster was reduced by 0.17 eV, and the dipole moment (D) was increased to 3.05 Debye. There was no noticeable change in the HOMO–LUMO energy gap. In all of the adsorption modes, it was found that the HOMO–LUMO energy gaps of the $CHCl_3$/ZnO complexes were in the range of 4.00–4.03 eV. Consequently, the adsorption of $CHCl_3$ on the ZnO nanocage had no significant effect on the HOMO–LUMO energy gap.

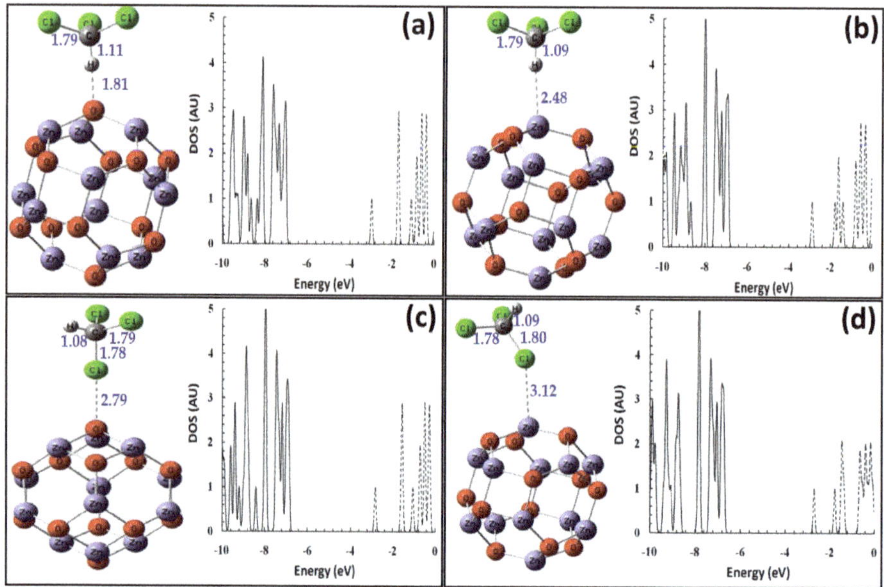

Figure 2. Optimized structures and DOSs of the $CHCl_3$ molecule adsorption on the $Zn_{12}O_{12}$ nanocage. (**a**) adsorption mode a, (**b**) adsorption mode b, (**c**) adsorption mode c, and (**d**) adsorption mode d. The distances are in Å, and the DOS is in arbitrary units. The solid and dashed lines represent occupied and virtual states, respectively.

Table 1. Electronic properties of the isomeric configurations of the $CHCl_3/Zn_{12}O_{12}$ complexes, namely: adsorption energy (E_{ads}; eV), HOMO (eV), LUMO (eV), Fermi level (E_{FL}; eV), HUMO–LUMO energy gap (E_g; eV), natural bond orbital (NBO) charge (Q; au), and dipole moment (D; Debye).

System	Bare $Z_{12}O_{12}$	(a)	(b)	(c)	(d)
E_{ads}	–	−0.38	0.07	−0.09	−0.04
E_{HOMO}	−6.81	−6.98	−6.90	−6.84	−6.71
E_{LUMO}	−2.79	−2.96	−2.89	−2.82	−2.71
E_{FL}	−4.80	−4.97	−4.90	−4.83	−4.71
E_g	4.02	4.03	4.00	4.02	4.00
Q_{CHCl3}	–	−0.02	0.00	−0.01	0.02
D	0.00	3.05	1.41	1.33	2.17

Additionally, to investigate the effect of the $CHCl_3$ concentration on the adsorption over the $Zn_{12}O_{12}$ nanocage, we performed geometric optimizations for n $CHCl_3$ molecules (n = 1, 2, 3, and 4) adsorbed simultaneously over the $Zn_{12}O_{12}$ nanocage to form $(CHCl_3)_n$/ZnO complexes. All of the $CHCl_3$ molecules had an orientation in which the H head of the $CHCl_3$ molecule was directed toward an O site of the $Zn_{12}O_{12}$ nanocage, which is the most energetic stable orientation, as presented in Figure 2. The adsorption energies (E_{ads}) were calculated using Equation (1), and are presented in Table 2.

Table 2. Electronic properties of the $(CHCl_3)_n/Zn_{12}O_{12}$ complexes, namely: adsorption energy (E_{ads}; eV), HOMO (eV), LUMO (eV), Fermi level (E_{FL}; eV), HUMO–LUMO energy gap (E_g; eV), NBO charge (Q; au), and dipole moment (D; Debye).

System	(a)	(b)	(c)	(d)
	n = 1	n = 2	n = 3	n = 4
E_{ads}	−0.38	−0.38	−0.60	−0.76
E_{HOMO}	−6.98	−7.14	−7.19	−7.39
E_{LUMO}	−2.96	−3.10	−3.14	−3.32
E_{FL}	−4.97	−5.12	−5.16	−5.36
E_g	4.03	4.04	4.05	4.07
Q_{CHCl3}	−0.02	−0.02	−0.02	−0.01
D	3.05	0.46	1.18	1.64

The optimized structures of $(CHCl_3)_n$/ZnO and their DOSs are shown in Figure 3. As indicated by Table 2, after the second molecule was adsorbed, the adsorption energy (E_{ads}) increased as n—the number of adsorbed $CHCl_3$ molecules—increased. Additionally, as the number of adsorbed $CHCl_3$ molecules increased, the Fermi level decreased. Furthermore, although there were no significant changes in the average acquired charge (Q_{CHCl_3}) on the $CHCl_3$ molecules, the dipole moment was sensitive to the number of adsorbed $CHCl_3$ molecules. The HOMO–LUMO energy gap (E_g), compared with that of the pristine $Zn_{12}O_{12}$ nanocage (4.02 eV), was not affected by the number of adsorbed $CHCl_3$ molecules.

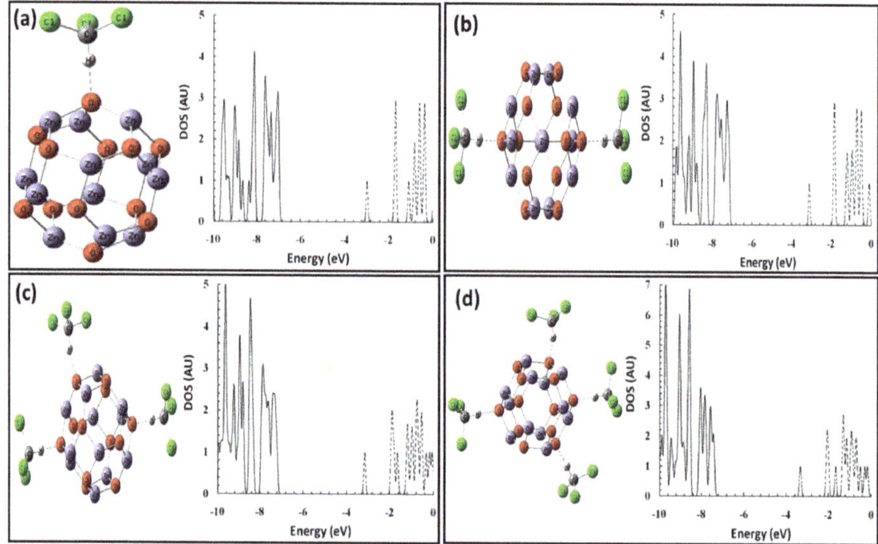

Figure 3. Optimized structures and DOSs for the $(CHCl_3)_n/Zn_{12}O_{12}$ nanocage. (**a**) $CHCl_3/Zn_{12}O_{12}$, (**b**) $(CHCl_3)_2/Zn_{12}O_{12}$, (**c**) $(CHCl_3)_3/Zn_{12}O_{12}$, and (**d**) $(CHCl_3)_4/Zn_{12}O_{12}$. The distances are in Å, and the DOS is in arbitrary units. The solid and dashed lines represent the occupied and virtual states, respectively.

3.3. O Atom Interaction with the $Zn_{12}O_{12}$ Nanocage

To improve the sensitivity of $Zn_{12}O_{12}$ to the $CHCl_3$ molecules, an O atom was deposited onto the cluster. To investigate the ability of the $Zn_{12}O_{12}$ nanocage to adsorb an O atom, the O atom was added at three different sites, namely: an O site, a Zn site, and the middle of the ZnO bond. Then, a full geometric optimization was performed for the $O/Zn_{12}O_{12}$ complexes. With the optimization, there are only two possible $O/Zn_{12}O_{12}$ complexes, as shown in Figure 4. The E_{ads} were calculated using Equation (2). As shown in Table 3, the E_{ads} values of the O atom on the $Zn_{12}O_{12}$ nanocage were −1.98 and −1.62 eV for complexes (a) and (b), respectively. This indicated that a chemical bond was formed between the O atom and the $Zn_{12}O_{12}$ cluster. Additionally, the NBO analysis indicated that the O atom gained negative charges (Q_O) of −0.71|e| and −0.61|e| for complexes (a) and (b), respectively.

Table 3. Electronic properties of the isomeric configurations of the $O/Zn_{12}O_{12}$ complexes, namely: adsorption energy (E_{ads}; eV), HOMO (eV), LUMO (eV), Fermi level (E_{FL}; eV), HUMO–LUMO energy gap (E_g; eV), NBO charge (Q; au), and dipole moment (D; Debye).

System	$O/Zn_{12}O_{12}$	$O/Zn_{12}O_{12}$
	(a)	(b)
E_{ads}	−1.98	−1.62
E_{HOMO}	−6.32	−6.57
E_{LUMO}	−2.79	−2.79
E_{FL}	−4.56	−4.68
E_g	3.53	3.78
Q_O	−0.71	−0.61
D	2.03	0.72

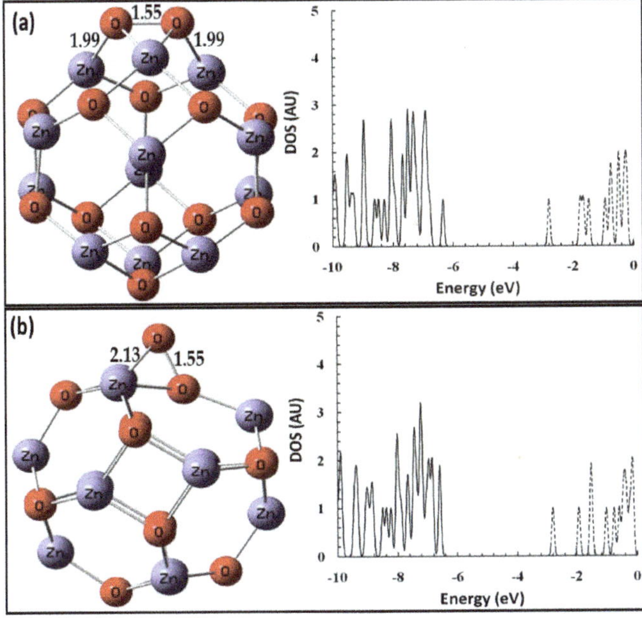

Figure 4. Optimized structures and DOSs of O atom adsorption on the $Zn_{12}O_{12}$ nanocage. (**a**) complex a, and (**b**) complex b. The distances are in Å, and the DOS is in arbitrary units. The solid and dashed lines represent the occupied and virtual states, respectively.

This strong interaction is attributed to the charge transfer from the $Zn_{12}O_{12}$ nanocage to the adsorbed O atom. As indicated by the DOS in Figure 4, the HOMO–LUMO energy gaps (E_g) of $O/Zn_{12}O_{12}$ for complexes (a) and (b) were reduced (to 3.53 and 3.78 eV, respectively) compared with that of the pristine $Zn_{12}O_{12}$ nanocage (4.02 eV; Table 1). Furthermore, for $O/Zn_{12}O_{12}$ complexes (a) and (b), increases of 0.24 and 0.12 eV, respectively, were observed for the Fermi level (E_{FL}), and the dipole moment increased to 2.03 and 0.72, respectively. This indicated that the deposited O atom significantly affected the electronic properties of the $Zn_{12}O_{12}$ nanocage, and consequently may have affected its ability to adsorb $CHCl_3$ molecules.

3.4. CHCl₃ Interaction with O Atoms Deposited on the Zn₁₂O₁₂ Nanocage

The $CHCl_3$ molecule could interact via its H head or Cl head, and the O atom could be deposited on the Zn or O sites of the nanocage; thus, there were four possible geometric structures for the $CHCl_3/O/Zn_{12}O_{12}$ complexes. Consequently, we performed geometric optimization for the four aforementioned $CHCl_3/O/Zn_{12}O_{12}$ complexes. During the optimization process, we found only three stable $CHCl_3/O/Zn_{12}O_{12}$ complexes, as shown in Figure 5. The properties of the interaction among the $CHCl_3$ molecule, deposited O atom, and $Zn_{12}O_{12}$ nanocage are presented in Table 4.

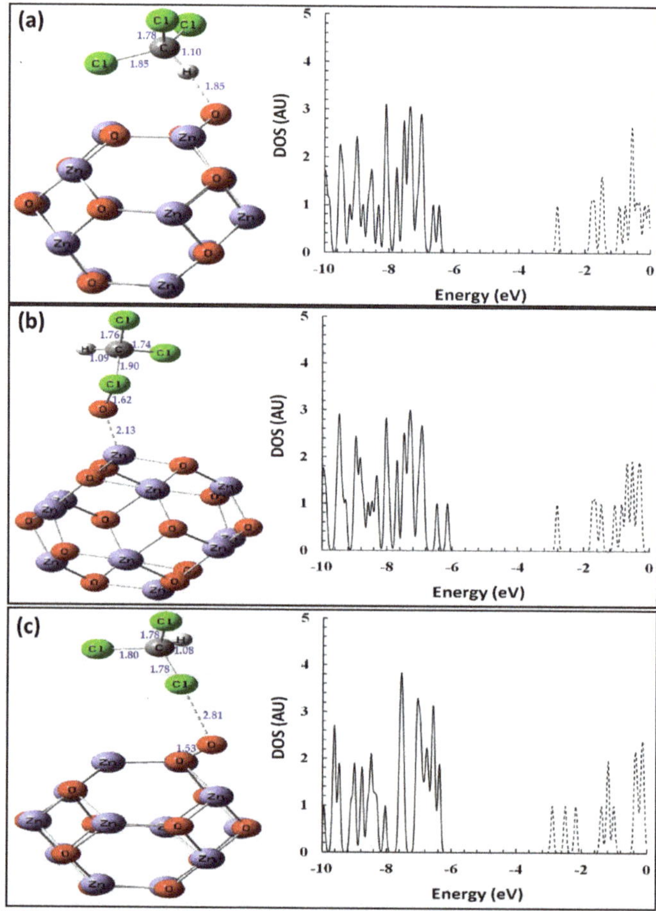

Figure 5. Optimized structures and DOSs for the $CHCl_3/O/Zn_{12}O_{12}$ nanocage. (**a**) complex a, (**b**) complex b, and (**c**) complex c. The distances are in Å, and the DOS is in arbitrary units. The solid and dashed lines represent the occupied and virtual states, respectively.

Table 4. Electronic properties of the isomeric configurations of the $CHCl_3/O/Zn_{12}O_{12}$ complexes, namely: adsorption energy (E_{ads}; eV), binding energy (E_b; eV), HOMO (eV), LUMO (eV), Fermi level (E_{FL}; eV), HUMO–LUMO energy gap (E_g; eV), NBO charge (Q; au).

System	(a)	(b)	(c)
E_{ads}	−2.44	−1.98	−0.92
E_b	−0.68	−0.15	−2.46
E_{HOMO}	−6.45	−6.13	−6.30
E_{LUMO}	−2.81	−2.85	−2.98
E_{FL}	−4.63	−4.49	−4.64
E_g	3.64	3.27	3.32
Q_{CHCl3}	0.07	0.03	0.86
Q_O	−0.63	−0.62	−0.76
Q_{ZnO}	0.56	0.59	−0.10
D	1.92	2.35	8.26

The adsorption energies (E_{ads}) for the complexes ranged from −0.92 to −2.44 eV. These values indicate a chemical interaction, which may have been due to a charge transfer. This can be explained by the NBO analysis, which revealed that in complexes (a) and (b), the deposited O atom gained negative charges of −0.63|e| and −0.62|e|, respectively. These charges were mainly transferred from the $Zn_{12}O_{12}$ nanocage, which gained positive charges of 0.56|e| and 0.59|e|, respectively. Additionally, there was a small charge from the $CHCl_3$ molecule, which gained positive charges of 0.07|e| and 0.03|e|, respectively. However, in complex (c), the charge was transferred from the $CHCl_3$ molecule, which gained a positive charge of 0.86|e|, to both the $Zn_{12}O_{12}$ nanocage and the deposited O atom, which gained negative charges of −0.10|e| and −0.76|e|, respectively. Clearly, the nature of the interaction in complex (c) was significantly different from those for complexes (a) and (b). This led to different binding energies between the $CHCl_3$ fragment and the $O/Zn_{12}O_{12}$ fragment of the $CHCl_3/O/Zn_{12}O_{12}$ complexes, which were −0.68, −0.15, and −2.46 eV for complexes (a), (b), and (c), respectively. Such interactions between $CHCl_3$ and the $O/Zn_{12}O_{12}$ nanocage led to an increase in the Fermi level (E_{FL}), from −4.64 to −4.49 eV, as well as a reduction of the HOMO–LUMO energy gaps (E_g), from 3.27 to 3.64 eV for the $CHCl_3/O/Zn_{12}O_{12}$ complexes, compared with 4.02 eV for the pristine $Zn_{12}O_{12}$ nanocage.

To examine the effect of the $CHCl_3$ concentration on the interaction with the $O/Zn_{12}O_{12}$ nanocage, we performed geometric optimizations for n $CHCl_3$ molecules (n = 1, 2, 3, and 4), adsorbed simultaneously over n deposited O atoms on the $Zn_{12}O_{12}$ nanocage. Each $CHCl_3$ molecule interacted via its Cl head with a deposited O atom on the Zn site of the $Zn_{12}O_{12}$ nanocage. This orientation yielded the highest binding energy between the $CHCl_3$ molecule and $O/Zn_{12}O_{12}$. The interaction energies were calculated using Equation (3). The optimized structures of $(CHCl_3)_n/O/Zn_{12}O_{12}$ and their DOSs are shown in Figure 6.

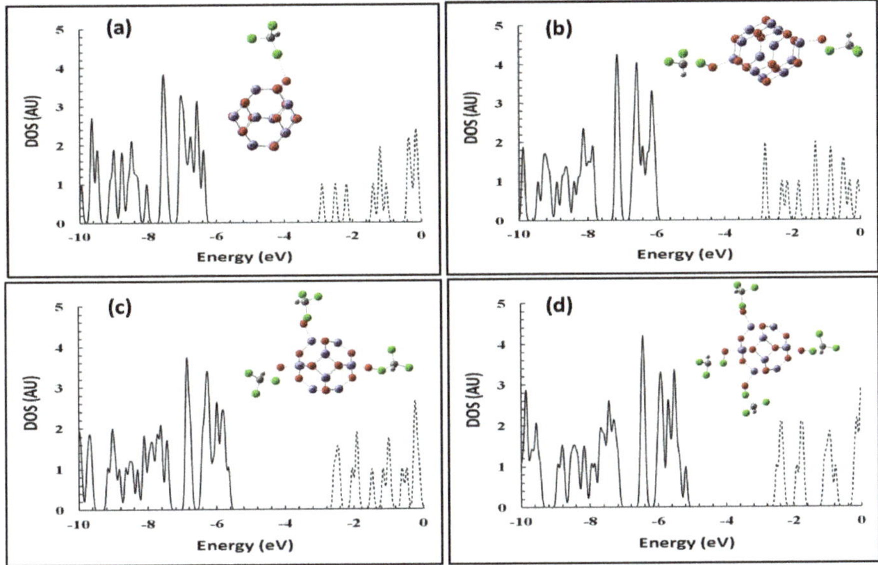

Figure 6. Optimized structures and DOSs for the $(CHCl_3)_n/O_n/Zn_{12}O_{12}$ nanocage. (a) $CHCl_3/O/Zn_{12}O_{12}$, (b) $(CHCl_3)_2/O_2/Zn_{12}O_{12}$, (c) $(CHCl_3)_3/O_3/Zn_{12}O_{12}$, and (d) $(CHCl_3)_4/O_4/Zn_{12}O_{12}$. The distances are in Å, and the DOS is in arbitrary units. The solid and dashed lines represent the occupied and virtual states, respectively.

The interaction energies are presented in Table 5. The adsorption energy remained relatively constant (approximately −0.96 eV) for the first three $CHCl_3$ interacting molecules, and decreased for the fourth $CHCl_3$ molecule (to −0.86 eV). Furthermore, as the number of adsorbed $CHCl_3$ molecules increased, the average acquired positive charges on $CHCl_3$ (Q_{CHCl_3}) decreased, and the negativity of the average charges on the deposited O atom (Q_O) decreased, while the negativity of the charges on the $Zn_{12}O_{12}$ nanocage increased. Additionally, with the increasing number of adsorbed $CHCl_3$ molecules, the Fermi level (E_{FL}) increased and the HOMO–LUMO energy gap (E_g) decreased, compared with the pristine $Zn_{12}O_{12}$ nanocages. The dipole moment of $(CHCl_3)_n/O/Zn_{12}O_{12}$ was sensitive to the number of $CHCl_3$ molecules.

Table 5. Electronic properties of the $(CHCl_3)_n/(O)_n/Zn_{12}O_{12}$ complexes, namely: adsorption energy (E_{ads}; eV), HOMO (eV), LUMO (eV), Fermi level (E_{FL}; eV), HUMO–LUMO energy gap (E_g; eV), NBO charge (Q; au), and dipole moment (D; Debye).

System	(a)	(b)	(c)	(d)
	$n = 1$	$n = 2$	$n = 3$	$n = 4$
E_{ads}	−0.92	−0.96	−0.96	−0.86
E_{HOMO}	−6.30	−6.03	−5.65	−5.19
E_{LUMO}	−2.98	−2.81	−2.57	−2.51
E_{FL}	−4.64	−4.42	−4.11	−3.85
E_g	3.32	3.22	3.08	2.68
Q_{CHCl3}	0.86	0.77	0.74	0.74
Q_O	−0.76	−0.61	−0.59	−0.58
Q_{ZnO}	−0.10	−0.33	−0.47	−0.62
D	8.26	3.26	4.15	6.45

3.5. $Zn_{12}O_{12}$ Nanocage as a Sensor for $CHCl_3$

It has been observed that during the adsorption process, the change in the HOMO–LUMO energy gap (E_g) is related to the sensitivity of the sorbent for the adsorbate. However, the reduction of E_g of the cluster significantly affects the electrical conductivity, as indicated by the following equation [52]:

$$\sigma \propto e^{(-E_g/2KT)} \tag{5}$$

where σ represents the electrical conductivity, K represents Boltzmann's constant, and T represents the temperature. According to Equation (5) and the E_g values in Tables 1 and 2, the adsorption of the $CHCl_3$ molecule in the gas phase did not lead to significant changes in the E_g of the $Zn_{12}O_{12}$ nanocage. According to Tables 4 and 5, the $CHCl_3$ molecule adsorption over the oxygenated ZnO significantly reduced the E_g values.

The energy difference between the nucleophile HOMO and electrophile LUMO is one of the important factors for HOMO–LUMO interactions. As previously mentioned, the chemical bonding between $CHCl_3$ and the oxygenated ZnO cluster in the $CHCl_3/O/Zn_{12}O_{12}$ complexes is due to the charge-transfer mechanism. It can be explained as the contribution from the HOMO of the $O/Zn_{12}O_{12}$ cluster to the vacant LUMO of the $CHCl_3$ molecule. Figure 7 shows the surfaces of the frontier molecular orbitals (FMOs; HOMO/LUMO) for $CHCl_3$, $Zn_{12}O_{12}$, $O/Zn_{12}O_{12}$, and $CHCl_3/O/Zn_{12}O_{12}$. The HOMO and the LUMO of the $Zn_{12}O_{12}$ cluster are localized on the Zn and O sites, respectively. Thus, the Zn sites are electrophilic centers, whereas the O sites are nucleophilic centers. This explains why the H atom of $CHCl_3$ is attached to the O site in the most stable structure of the $CHCl_3/Zn_{12}O_{12}$ complex. Additionally, the HOMO of the $O/Zn_{12}O_{12}$ cluster is localized around the deposited O atom. This explains why $CHCl_3$ is attracted to the deposited atom of the $O/Zn_{12}O_{12}$ cluster.

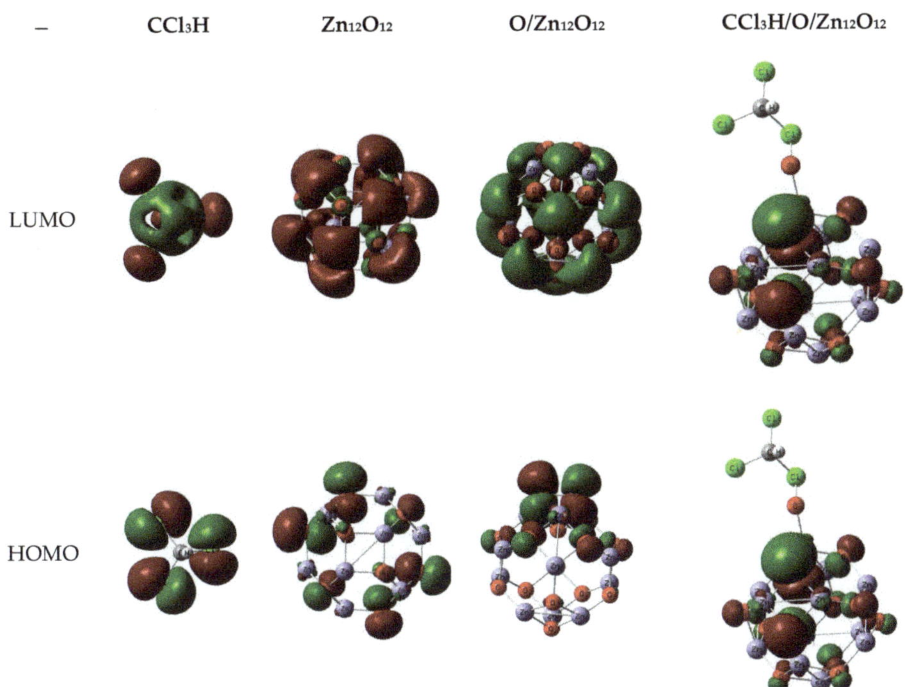

Figure 7. Frontier molecular orbital (FMO) surfaces (HOMO–LUMO) for $CHCl_3$, $Zn_{12}O_{12}$, $O/Zn_{12}O_{12}$, and $CHCl_3/O/Zn_{12}O_{12}$.

Figure 8 shows the energy diagrams of the FMOs (HOMO/LUMO) for $CHCl_3$, $Zn_{12}O_{12}$, $O/Zn_{12}O_{12}$, and $CHCl_3/O/Zn_{12}O_{12}$. Our FMO studies revealed that the deposited O atom increased the HOMO of the ZnO cluster from -6.81 to -6.32 eV. Consequently, the energy gap between the HOMO of ZnO and the LUMO of $CHCl_3$ decreased, making the charge transfer from the $O/Zn_{12}O_{12}$ cluster to the $CHCl_3$ easier than that from the pristine $Zn_{12}O_{12}$ cluster. Thus, the $O/Zn_{12}O_{12}$ cluster is more sensitive to the $CHCl_3$ molecule than the pristine $Zn_{12}O_{12}$ cluster.

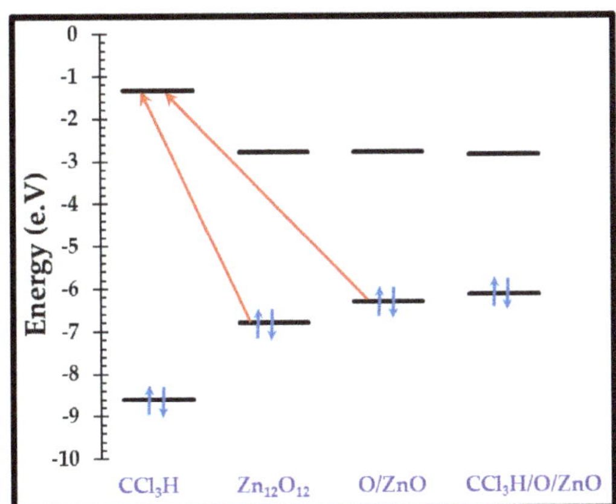

Figure 8. Energy diagram of the FMOs (HOMO/LUMO) for $CHCl_3$, $Zn_{12}O_{12}$, $O/Zn_{12}O_{12}$, and $CHCl_3/O/Zn_{12}O_{12}$.

4. Conclusions

A DOS study of chloroform sensing, based on ZnO nanocrystals, was performed via calculations implemented using the Gaussian 09 suite of programs. A geometric optimization was performed for the ZnO nanocrystal. The DOS for the ZnO nanocrystal was calculated. The calculated gap between the HOMO and the LUMO was found to be 4.02 eV. Furthermore, the effect of the concentration of $CHCl_3$ on its adsorption over the ZnO nanocrystals was investigated. The results indicated that the electrical properties of ZnO were not affected by the concentration of $CHCl_3$. Additionally, the effect of depositing O atoms on the ZnO adsorption properties was examined. The results indicated that the adsorption of $CHCl_3$ on the oxygenated ZnO reduced its bandgap. The findings of this study confirm that the deposition of O on a ZnO nanocluster increases its sensitivity to $CHCl_3$, and may facilitate $CHCl_3$ removal or detection.

Author Contributions: Conception and design of the experiments, H.Y.A., H.M.B. and A.U.; implementation of the experiments, H.Y.A., H.M.B. and A.U.; analysis of the data, contribution of the analysis tools, and writing and revision of the paper, H.Y.A., H.M.B., A.U., H.F. and O.Y.A.

Funding: This work is funded by Deanship of Scientific Research at King Saud University under the research groups grant (No. RG-1435-052).

Acknowledgments: The authors would like to extend their sincere appreciation to the Deanship of Scientific Research at King Saud University for funding this research group (No. RG-1435-052). The authors thank RSSU at King Saud University for their technical support.

Conflicts of Interest: The authors declare no conflict of interest.

References

1. Batzill, M.; Diebold, U. The surface and materials science of tin oxide. *Prog. Surf. Sci.* **2005**, *79*, 47–154. [CrossRef]
2. Wang, Z.-H.; Yang, C.-C.; Yu, H.-C.; Peng, Y.-M.; Su, Y.-K. Electron emission enhanced properties of gold nanoparticle-decorated ZnO nanosheets grown at room temperature. *Sci. Adv. Mater.* **2018**, *10*, 1675–1679. [CrossRef]
3. Kim, E.-B.; Lee, J.-E.; Akhtar, M.S.; Ameen, S.; Fijahi, L.; Seo, H.-K.; Shin, H.-S. Electrical sensor based on hollow ZnO spheres for hydrazine detection. *J. Nanoelectron. Optoelectron.* **2018**, *13*, 1769–1776. [CrossRef]
4. Bader, C. The dangers of chloroform, etc., and the nitrite of amyl. *Lancet* **1875**, *105*, 644. [CrossRef]
5. Featherstone, P.J.; Ravindran, H. Townley's chloroform inhaler. *J. Anesth. Hist.* **2018**, *4*, 103–108. [CrossRef] [PubMed]
6. Verma, P.; Tordik, P.; Nosrat, A. Hazards of improper dispensary: Literature review and report of an accidental chloroform injection. *J. Endod.* **2018**, *44*, 1042–1047. [CrossRef]
7. Selby, R. Chloroform-Its benefits and its dangers. *Lancet* **1848**, *51*, 190. [CrossRef]
8. Zhang, K.; Zhang, H.; Liu, R.; Yang, Z.; Yuan, Y.; Wang, F. Fabrication of aluminum doped zinc oxide thin films with various microstructures for gas sensing application. *Sci. Adv. Mater.* **2018**, *10*, 367–372. [CrossRef]
9. Ahad, I.Z.M.; Harun, S.W.; Gan, S.N.; Phang, S.W. Polyaniline (PAni) optical sensor in chloroform detection. *Sens. Actuators B Chem.* **2018**, *261*, 97–105. [CrossRef]
10. Sheng, K.; Lu, H.; Sun, A.; Wang, Y.; Liu, Y.; Chen, F.; Bian, W.; Li, Y.; Kuang, R.; Sun, D. A naked-eye colorimetric sensor for chloroform. *Chin. Chem. Lett.* **2019**, *30*, 895–898. [CrossRef]
11. Sharma, S.; Nirkhe, C.; Pethkar, S.; Athawale, A.A. Chloroform vapour sensor based on copper/polyaniline nanocomposite. *Sens. Actuators B Chem.* **2002**, *85*, 131–136. [CrossRef]
12. Wang, K.; Li, S.; Jiang, Y.; Hu, M.; Zhai, Q.G. A pillar-layered metal-organic framework as luminescent sensor for selective and reversible response of chloroform. *J. Solid State Chem.* **2017**, *247*, 39–42. [CrossRef]
13. Zhang, K.; Zhang, H.; Sun, S.; Yang, Z.; Yuan, Y.; Wang, F. Enhanced gas sensing properties based on ZnO-decorated nickel oxide thin films for formaldehyde detection. *Sci. Adv. Mater.* **2018**, *10*, 373–378. [CrossRef]
14. Kumar, R.; Al-Dossary, O.; Kumar, G.; Umar, A. Zinc oxide nanostructures for NO_2 gas sensor applications: A review. *Nano Micro Lett.* **2015**, *7*, 97–120. [CrossRef] [PubMed]

15. Jia, R.; Xie, P.; Feng, Y.; Chen, Z.; Umar, A.; Wang, Y. Dipole-modified graphene with ultrahigh gas sensibility. *Appl. Surf. Sci.* **2018**, *440*, 409–414. [CrossRef]

16. Zhou, Q.; Xu, L.; Umar, A.; Chen, W.; Kumar, R. Pt nanoparticles decorated SnO_2 nanoneedles for efficient CO gas sensing applications. *Sens. Actuators B Chem.* **2018**, *256*, 656–664. [CrossRef]

17. Umar, A.; Alshahrani, A.A.; Algarni, H.; Kumar, R. CuO nanosheets as potential scaffolds for gas sensing applications. *Sens. Actuators B Chem.* **2017**, *250*, 24–31. [CrossRef]

18. Al-Hadeethi, Y.; Umar, A.; Ibrahim, A.A.; Al-Heniti, S.H.; Kumar, R.; Baskoutas, S.; Raffah, B.M. Synthesis, characterization and acetone gas sensing applications of Ag-doped ZnO nanoneedles. *Ceram. Int.* **2017**, *43*, 6765–6770. [CrossRef]

19. Umar, A.; Lee, J.-H.; Kumar, R.; Al-Dossary, O. Synthesis and characterization of CuO nanodisks for high-sensitive and selective ethanol gas sensor applications. *J. Nanosci. Nanotechnol.* **2017**, *17*, 1455–1459. [CrossRef]

20. Al-Hadeethi, Y.; Umar, A.; Al-Heniti, S.H.; Kumar, R.; Kim, S.H.; Zhang, X.; Raffah, B.M. 2D Sn-doped ZnO ultrathin nanosheet networks for enhanced acetone gas sensing application. *Ceram. Int.* **2017**, *43*, 2418–2423. [CrossRef]

21. Nam, G.; Leem, J.Y. A new technique for growing ZnO nanorods over large surface areas using graphene oxide and their application in ultraviolet sensors. *Sci. Adv. Mater.* **2018**, *10*, 405–409. [CrossRef]

22. Zhang, N.; Yu, K.; Li, Q.; Zhu, Z.Q.; Wan, Q.J. Room-temperature high-sensitivity H_2S gas sensor based on dendritic ZnO nanostructures with macroscale in appearance. *Appl. Phys.* **2008**, *103*, 104305. [CrossRef]

23. Jimenez, I.; Arbiol, J.; Dezanneau, G.; Cornet, A.; Morante, J.R. Crystalline structure, defects and gas sensor response to NO_2 and H_2S of tungsten trioxide nanopowders. *Sens. Actuators B Chem.* **2003**, *93*, 475–485. [CrossRef]

24. Chen, Z.; Wang, J.; Umar, A.; Wang, Y.; Li, H.; Zhou, G. Three-dimensional crumpled graphene-based nanosheets with ultrahigh NO_2 gas sensibility. *ACS Appl. Mater. Interfaces* **2017**, *9*, 11819–11827. [CrossRef]

25. Umar, A.; Akhtar, M.S.; Al-Assiri, M.S.; Al-Salami, A.E.; Kim, S.H. Composite CdO-ZnO hexagonal nanocones: Efficient materials for photovoltaic and sensing applications. *Ceram. Int.* **2018**, *44*, 5017–5024. [CrossRef]

26. Umar, A. *Encyclopedia of Semiconductor Nanotechnology*; American Scientific Publishers: Los Angeles, CA, USA, 2017.

27. Choi, H.J.; Lee, Y.M.; Boo, J.H. Well-aligned ZnO nanorods assisted with close-packed polystyrene monolayer for quartz crystal microbalance. *Sci. Adv. Mater.* **2018**, *10*, 610–615. [CrossRef]

28. Yang, J.; Yi, W.; Zhang, L.; Li, T.; Chao, Z.; Fan, J. Facile fabrication of ZnO nanomaterials and their photocatalytic activity study. *Sci. Adv. Mater.* **2018**, *10*, 1721–1728. [CrossRef]

29. Umar, A.; Hahn, Y.B. *Metal Oxide Nanostructures and Their Applications*; American Scientific Publishers: Los Angeles, CA, USA, 2010.

30. Wang, J.X.; Sun, X.W.; Yang, Y.; Huang, H.; Lee, Y.C.; Tan, O.K.; Vayssieres, L. Hydrothermally grown oriented ZnO nanorod arrays for gas sensing applications. *Nanotechnology* **2006**, *17*, 4995–4998. [CrossRef]

31. Paraguay, F.; Miki-Yoshida, M.; Morales, J.; Solis, J.; Estrada, L.W. Influence of Al, In, Cu, Fe and Sn dopants on the response of thin film ZnO gas sensor to ethanol vapour. *Thin Solid Films* **2000**, *373*, 137–140. [CrossRef]

32. Chaudhary, S.; Kaur, Y.; Umar, A.; Chaudhary, G.R. Ionic liquid and surfactant functionalized ZnO nanoadsorbent for recyclable proficient adsorption of toxic dyes from waste water. *J. Mol. Liq.* **2016**, *224*, 1294–1304. [CrossRef]

33. Guo, R.; Cheng, X.; Gao, S.; Zhang, X.; Xu, Y.; Zhao, H.; Huo, L. Highly selective detection of saturated vapors of abused drugs by ZnO nanorod bundles gas sensor. *Appl. Surf. Sci.* **2019**, *485*, 266–273. [CrossRef]

34. Shao, C.; Chang, Y.; Long, Y. High performance of nanostructured ZnO film gas sensor at room temperature. *Sens. Actuators B Chem.* **2014**, *204*, 666–672. [CrossRef]

35. Ghenaatian, H.R.; Baei, M.T.; Hashemian, S. $Zn_{12}O_{12}$ nano-cage as a promising adsorbent for CS_2 capture. *Superlattices Microstruct.* **2013**, *58*, 198–204. [CrossRef]

36. Baei, M.T.; Tabar, M.B.; Hashemian, S. $Zn_{12}O_{12}$ fullerene-like cage as a potential sensor for SO_2 detection. *Adsorpt. Sci. Technol.* **2013**, *3*, 469–476. [CrossRef]

37. Ammar, H.Y. CH_2O adsorption on M (M = Li, Mg and Al) atom deposited ZnO nano-cage: DFT study. *Key Eng. Mater.* **2018**, *786*, 384. [CrossRef]

38. Mayya, D.S.; Prasad, P.; Kumar, J.R.N.; Aswanth, P.; Mayur, G.; Aakanksha, M.; Bindushree, N. Nanocrystalline zinc oxide thin film gas sensor for detection of hydrochloric acid, ethanolamine, and chloroform. *Int. J. Adv. Res. Sci. Eng. Technol.* **2018**, *7*, 809–818.

39. Frisch, M.J.; Trucks, G.W.; Schlegel, H.B.; Scuseria, G.E.; Robb, M.A.; Cheeseman, J.R.; Montgomery, J.A., Jr.; Vreven, T.; Kudin, K.N.; Burant, J.C.; et al. *Gaussian 09*; Gaussian, Inc.: Wallingford, CT, USA, 2004.

40. Becke, A.D. Density-functional thermochemistry. III. The role of exact exchange. *J. Chem. Phys.* **1993**, *98*, 5648. [CrossRef]

41. Vosko, S.H.; Wilk, L.; Nusair, M. Accurate spin-dependent electron liquid correlation energies for local spin density calculations: A critical analysis. *Can. J. Phys.* **1980**, *58*, 1200–1211. [CrossRef]

42. Becke, A.D. Density-functional exchange-energy approximation with correct asymptotic behavior. *Phys. Rev. A* **1988**, *38*, 3098. [CrossRef]

43. Lee, C.; Yang, W.; Parr, R.G. Development of the Colle-Salvetti correlation-energy formula into a functional of the electron density. *Phys. Rev. B* **1988**, *37*, 785. [CrossRef]

44. Miehlich, B.; Savin, A.; Stoll, H.; Preuss, H. Results obtained with the correlation energy density functionals of becke and Lee, Yang and Parr. *Chem. Phys. Lett.* **1989**, *157*, 200–206. [CrossRef]

45. Beheshtian, J.; Peyghan, A.A.; Bagheri, Z. Adsorption and dissociation of Cl_2 molecule on ZnO nanocluster. *Appl. Surf. Sci.* **2012**, *258*, 8171. [CrossRef]

46. Boyle, N.M.O.; Tenderholt, A.L.; Langner, K.M. Cclib: A library for package-independent computational chemistry algorithms. *J. Comp. Chem.* **2008**, *29*, 839. [CrossRef] [PubMed]

47. Glendening, E.D.; Reed, A.E.; Carpenter, J.E.; Weinhold, F. *NBO Program, Version 3.1*; University of Wisconsin: Madison, WI, USA, 2001.

48. Narayanaswamy, A.; Xu, H.F.; Pradhan, N.; Peng, X.G. Crystalline nanoflowers with different chemical compositions and physical properties grown by limited ligand protection. *Angew. Chem. Int. Ed.* **2006**, *45*, 5361. [CrossRef] [PubMed]

49. Joshi, P.; Shewale, V.; Pandey, R.; Shanker, V.; Hussain, S.; Karna, S.P. Site specific interaction between ZnO nanoparticles and tryptophan: A first principles quantum mechanical study. *Phys. Chem. Chem. Phys.* **2011**, *13*, 476. [CrossRef] [PubMed]

50. Shalabi, A.S.; Abdel Aal, S.; Kamel, M.A.; Taha, H.O.; Ammar, H.Y.; Abdel Halim, W.S. The role of oxidation states in F_{A1} Tl^{n+} ($n = 1, 3$) lasers and CO interactions at the (1 0 0) surface of NaCl: An ab initio study. *Chem. Phys.* **2006**, *328*, 8–16. [CrossRef]

51. Shalabi, A.S.; Abdel Aal, S.; Ammar, H.Y. Artificial polarization effects on F_{A1}:Sr^{2+} lasers and NO interactions at NaCl (001) surface: First principles calculations. *J. Mol. Struct. Theochem.* **2007**, *823*, 47–58. [CrossRef]

52. Li, S. *Semiconductor Physical Electronics*, 2nd ed.; Springer: New York, NY, USA, 2006.

Article

Multi-Layered Mesoporous TiO$_2$ Thin Films: Photoelectrodes with Improved Activity and Stability

Enno Gent, Dereje H. Taffa * and Michael Wark

Institute of Chemistry, Chemical Technology 1, Carl von Ossietzky University of Oldenburg, D-26129 Oldenburg, Germany; enno.gent@uol.de (E.G.); michael.wark@uol.de (M.W.)
* Correspondence: dereje.hailu.taffa@uol.de; Tel.: +49-441-798-3279

Received: 25 August 2019; Accepted: 24 September 2019; Published: 28 September 2019

Abstract: This work aims at the identification of porous titanium dioxide thin film (photo)electrodes that represent suitable host structures for a subsequent electrodeposition of plasmonic nanoparticles. Sufficient UV absorption and electrical conductivity were assured by adjusting film thickness and TiO$_2$ crystallinity. Films with up to 10 layers were prepared by an evaporation-induced self-assembly (EISA) method and layer-by-layer deposition. Activities were tested towards the photoelectrochemical oxidation of water under UV illumination. Enhanced activities with each additional layer were observed and explained with increased amounts of immobilized TiO$_2$ and access to more active sites as a combined effect of increased surface area, better crystallinity and improved transport properties. Furthermore, films display good electrochemical and mechanical stability, which was related to the controlled intermediate thermal annealing steps, making these materials a promising candidate for future electrochemical depositions of plasmonic noble metal nanoparticles that has been further demonstrated by incorporation of gold.

Keywords: titanium dioxide; mesoporous; thin film; multi-layered; photoanode; semiconductor; photoelectrochemical water oxidation

---

1. Introduction

Since the seminal discovery of water photolysis by Fujishima and Honda in 1972 [1], TiO$_2$-based materials have experienced a rarely diminished attention in material research. Due to its unique set of properties, namely being inexpensive, environmentally benign, stability in aqueous media, recyclability and having favorable band energies, TiO$_2$ is an appealing material choice [2–4]. Moreover, its very flexible processability and intensively explored synthesis routes to modify its structure lead to a wide range of energy and environmental applications [5–7]. To name a few non-photocatalytic applications, these can be sensors [8,9], photovoltaic devices [10,11] and Li-ion batteries [12,13]. Additionally, TiO$_2$ can be found in various (heterogenous) photocatalytic applications ranging from self-cleaning/sterilizing and anti-fogging surfaces [3,14], over water and air purification [15,16] and anticorrosive coatings [17,18] to CO$_2$ photoreduction [19–21] and H$_2$ evolution from water [5,22].

To achieve a good photocatalytic performance, it is necessary to know the desired application before optimizing the catalyst with respect to both its structure and its optoelectrochemical properties. Structurally, mass transfer (reactant to active sites), charge transfer (catalyst surface to reactant) and charge transport (catalyst bulk to its surface) must be considered [4]. Access to more active sites, an improved collection/harvesting of photogenerated charge carriers and light trapping [23] are desirable approaches towards higher quantum yields. In light of this, extensively documented experimental techniques towards various target structures have emerged over the course of recent decades [4]. Today, TiO$_2$ can be shaped into films [24–26], spheres [27,28], fibers [29,30], rods [31,32], tubes [26,33], etc. with variable pore sizes and/or pore size distributions ranging from micro- [34,35] to macroporous [36,37] and—occasionally—bimodal hierarchical [38,39].

Optoelectrochemical properties highly depend on the nanostructural features of titanium dioxide, such as crystallinity and crystallite size, grain boundaries and the types and amounts of defects. However, there are also two major drawbacks that intrinsically impair the efficiency of pure TiO_2-based photocatalysts: the first is a band gap of $\approx$3.0–3.2 eV which limits photon absorption to the UV spectrum and thus to the small fraction of $\approx$5% solar photon energy [2,40]; the second is relatively fast recombination rates of photogenerated electron–hole pairs due to their limited mobility and short lifetimes [2,41]. There are numerous approaches to address these two issues including metal and/or nonmetal-doping [40,42,43] (to partially replace either the Ti^{4+} and/or O^{2-} sites), heterojunction formation [20,44], Z-schemes [45,46], decoration with quantum dots [47] and—with more recent success—hydrogenation towards "black" TiO_2 [48] or introduction of surface plasmon resonance (SPR) active noble metal nanoparticles (NPs) [41,49–53]. To develop TiO_2 photocatalysts, it is, therefore, necessary to address both origins—structure and intrinsic optoelectrochemical properties.

The scope of the present study is focused on a TiO_2-based system to identify suitable mesoporous host structures for further modifications related to both origins. It is important to understand the intended modifications, since they define requirements for the investigated material. At a later stage of this study, the electrochemical deposition (ED) of SPR-active Au-NPs is outlined. The basic appeal of decorating TiO_2 with such NPs is that both major drawbacks can be tackled simultaneously: Visible-light response can be attained via the SPR effect plus subsequent injection of hot electrons into the conduction band of TiO_2 [54] (where they can be consumed for reduction processes [55,56]) and improved electron/hole separation, since the metal–TiO_2 interface forms a Schottky barrier thus decelerating charge carrier recombination [41,51]. SPR features are not only size and shape dependent [57], but the surrounding chemical environment and its interaction with the NPs also plays a crucial role. When considering the latter fact together with the intended ED, the most important requirement for the TiO_2 host system becomes apparent: being a (meso)porous TiO_2 electrode.

Mesoporous TiO_2 thin films (MTTFs) not only display tunable host matrices due to their well-definable pore system [58,59], they are also beneficial for facilitated product recovery due to their immobilized nature. A frequently reported technique towards MTTFs—which is commonly used for dip-, spin-coating [60]—is the balanced combination of sol-gel chemistry and the evaporation-induced self-assembly (EISA) in the presence of surfactants which yield the metal oxide and the porous network, respectively [61–63]. Five steps, namely (i) precursor selection and initial sol preparation, (ii) deposition, (iii) controlled aging, (iv) template removal and (v) wall crystallization, govern the final film properties as Soler-Illia et al. pointed out [62]. Recently, the same authors contributed to state-of-the-art designs of plasmon-enhanced MTTFs with structural control of both the host and SPR-active NPs guests [64–68]. Their groups prepared well-ordered MTTFs derived either from mild oxidative at 350 °C air calcination [64–66] or extractive template removal [67,68] and employed an impregnation–reduction method to introduce noble metal NPs. The resulting films were for example tested for surface-enhanced Raman scattering (SERS) activity for sensing applications.

The electrochemical incorporation of metallic NPs into MTTFs necessitates electronic conductivity of the films. Thus, improved crystallinity is required to allow deposition throughout the film rather than on the electrode's back contact. This can be achieved by elevated annealing temperatures which usually are accompanied by significant mesostructural deterioration due to mass migration and crystal growth [58,62]. The trade-off between loss of active surface area and improved crystallinity can be manipulated to some extent with an additional thermal treatment under inert-gas atmosphere [47,69]. However, the degree of mesoporosity—which determines accessibility of active sites and diffusion of molecules through the film—seems to be more important than the order of the porosity [62,70].

Photocatalytic activity not only depends on crystal sizes and structure alone but also on the portion of absorbed photons. The EISA approach mostly results limited amount of material per deposited layer [71–73] due the low viscosity and diluted concentrations of EISA-sols [58]. This necessitates a more time-consuming but worthwhile layer-by-layer deposition to prepare thicker films [47,72,74,75]. Such MTTFs display considerably high active surface sites that are accessible throughout the film as result of its mesoporosity and maximize the number of exploitable photogenerated electron–hole pairs for photocatalytic reactions.

Briefly, the present study aims at the preparation and identification of MTTF (photo)electrode materials that represent suitable host structures for a subsequent (photo)electrodeposition of plasmonic noble metal nanoparticles. Thus, the layer-by-layer deposition of multi-layered TiO_2 onto FTO/glass substrates and its effect on performance, structure and film stability is investigated. At first, single-layered MTTFs were prepared according to an established procedure and were further optimized with respect to a calcination temperature of 550 °C. In the second part, the layer-by-layer deposition and characterization of multi-layered films are performed showing that an undiminished activity up to 10 layers is observed. Thirdly, performance and mechanical stability tests are conducted to demonstrate the film's durability. Finally, their suitability as porous hosts for subsequent loadings is briefly demonstrated in terms of pulsed electrochemical incorporation of SPR-active Au-NPs.

2. Materials and Methods

2.1. Chemicals, Materials

FTO/glass substrates (Pilkington "NSG TEC^{TM} A7", 7 $\Omega\square^{-1}$, 2.2 mm thickness, manufacturer: NSG Group, Gladbeck, Germany), acetone (Carl Roth, $\geq$99.9%), Pluronic$^{©}$ P123 (Sigma Aldrich, M $\approx$ 5800 g·mol^{-1}), glycerol (VWR, 99.8%), 1,6-diisocyanatohexane (Sigma Aldrich, $\geq$98%), abs. ethanol (Fisher Scientific, 99.5%), HCl(aq) (Fisher Scientific, 37 wt.%), Ti(OEt)$_4$ (Alfa Aesar, >99%), de-ionized H_2O (18.2 MΩ cm), HAuCl$_4$ × 3 H_2O (Alfa Aesar, 48.5–50.25 wt.% Au), HClO$_4$ (Bernd Kraft, 70 wt.%). All chemicals were used without further purification.

2.2. Preparation of Mesoporous TiO_2 Thin Films

The preparation of MTTFs proceeded according to the flowchart depicted in Figure 1a.

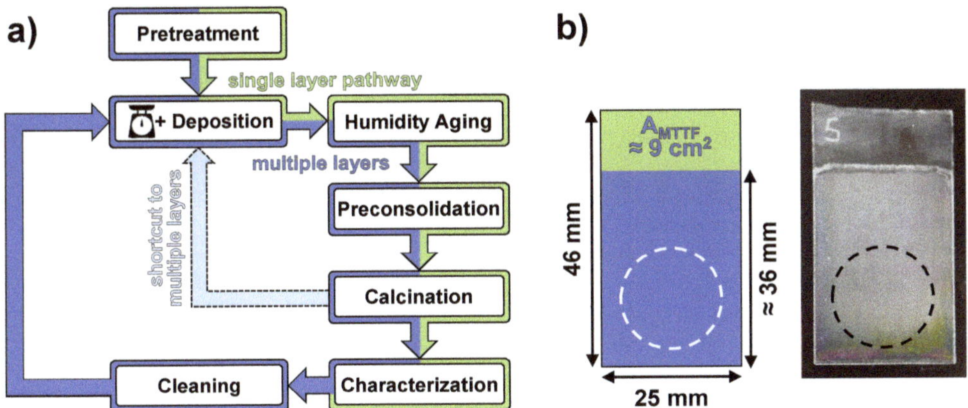

Figure 1. (**a**) Flowchart of steps involved in the film preparation. Green and blue arrows represent pathways to single- and multi-layered MTTFs, respectively. The pale blue arrow indicates a shortcut by skipping characterization/cleaning of MTTFs. (**b**) Scheme and photograph of a TiO_2 thin film. Green and blue areas represent the bare FTO and TiO_2 part, respectively. The circle indicates the region where electrodeposition might take place later; but more importantly, where optical spectroscopy and photoelectrochemistry are performed.

Pretreatment: FTO/glass substrates were cut into pieces of 25 mm × 46 mm and cleaned by four consecutive ultrasonication steps (namely in 0.1 M NaOH(aq), 0.1 M HCl(aq), ethanol and acetone; each lasting 10 min at 37 kHz). Cleaned substrates were modified with a so-called "cross-linked P123".

This strategy has proven to promote the orthogonal orientation of two-dimensional mesochannels (by providing a surface chemically neutral towards both the hydrophilic (PEO) and hydrophobic (PPO) parts of the triblock-copolymer of P123) and was first demonstrated for MTTFs by the Rankin group [76,77]. For doing so, cleaned FTO/glass slides were dip-coated into an acetone-based solution with equimolar amounts of P123 and 1,6-diisocyanatohexane (both 0.696 mM) where the triol glycerol enabled cross-linkage. This solution was prepared by dissolving 0.404 g P123 and a droplet of glycerol into 100 mL acetone (15 min ultrasonication at 37 kHz), precooling in a refrigerator and addition of 12 mg of the diisocyanate inside a glovebox. It was readily used due to the ongoing isocyanate-hydroxyl polyaddition reaction towards polyurethane and the dip-coating procedure was performed at ambient conditions with a 70 mm·min^{-1} immersion rate, 20 s holding time and withdrawal rate of 20 mm·min^{-1} with an *IDLAB 4 AC Coater* device.

Deposition (dip-coating): EISA-sols were prepared by dissolving 1.30 g P123 in 15.0 g ethanol (15 min ultrasonication at 37 kHz). After being cooled in the refrigerator for 30 min, 3.06 g of concentrated HCl(aq) were added dropwise under vigorous stirring. The resulting mixture was further ultrasonicated and placed in a refrigerator for 30 min again. The precooled solution was transferred into a glovebox where 4.20 g of Ti(OEt)$_4$ were added dropwise under vigorous stirring. The resulting molar ratio in the EISA-sol was Ti(OEt)$_4$:P123:EtOH:HCl:H$_2$O = 1.00:0.0122:17.7:1.6:5.9. Both the chemicals and their stoichometric ratios were adopted from Choi et al. and are known to promote the formation of hexagonal mesopores [78]. However, little changes regarding the used solvent were made towards thicker films based on the procedure described in [76].

Films were deposited onto cross-linked FTO/glass substrates via dip-coating (*IDLAB 4 AC Coater*) into a teflon-made tank (reservoir dimensions: 38 mm height, 28 mm width, 8 mm depth) with a 70 mm·min^{-1} immersion rate, 20 s holding time and a 20 mm·min^{-1} withdrawal rate. Air-conditioning allowed operation at T < 15 °C and relative humidities >70 %RH. After 2 min of hanging inside the dip-coater, the substrates were taken out for the aging phase. Prior to the deposition, the backsides of the cross-linked substrates were masked with a solvent resistant scotch tape.

Humidity aging, preconsolidation, calcination: As-deposited films were aged at low temperatures (4 °C) and high relative humidities (>90%) for 2–3 h. This environment has proven to favor the relatively quick formation of ordered mesostructures [78] and was established by placing substrates in dedicated aging boxes, namely sealable boxes with a reservoir of saturated KNO$_3$(aq) that can be stored inside a refrigerator. To evaporate remaining volatile compounds and complete the polycondensation, the films were stored inside a drying furnace at 100 °C overnight. For template removal and network crystallization, the scotch tape was peeled of and preconsolidated films were calcined in air by increasing the temperature with +5 °C min^{-1} to elevated values of 350–650 °C holding for 2 h and natural cooling to room temperature.

Pulsed electrodeposition of gold nanoparticles was carried out in a tree-electrode configuration schematically depicted in Figure 2a. A Pt coil, Ag/AgCl(sat. KCl) and a MTTF were connected to an *AMEL 7050* potentiostat and were used as counter, reference and working electrodes, respectively. A salt bridge was used to separate reference electrode and deposition bath.

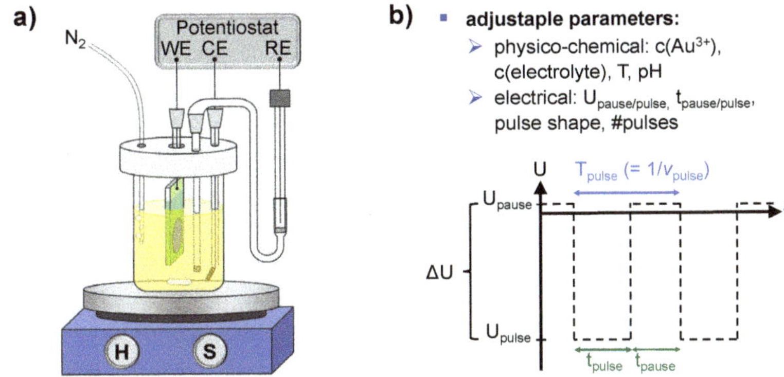

Figure 2. Pulsed electrodeposition of Au-NPs into MTTFs. (**a**) schematic representation of the used setup and (**b**) adjustable parameters.

An aqueous solution of 10 mM $HClO_4$ and 70 µM $HAuCl_4$ with a pH value of 2.1 was used as deposition bath and prior to the addition of the gold salt, the 10 mM $HClO_4$(aq) was purged with nitrogen for 30 min [50]. To ensure that a defined geometric sample area is exposed to the electrolyte solution, the MTTF was masked with a solvent resistant scotch tape leaving out a circular zone of 2 cm diameter (=3.14 cm²). Electrodeposition was carried out at room temperature on single-layered MTTFs and involved reduction reaction towards elemental gold shown in Equation (1) [79]. Two oxidation reactions at the Pt anode are expectable when considering the used chemicals and applied potentials. They are mentioned in Equations (2) and (3) [79] but are of minor importance here.

The used electrochemical parameters rely on a previous study by Bannat et al. and involved a pulse potential of U_{pulse} = −2.0 V vs. Ag/AgCl, a pause potential of U_{pause} = +0.1 V vs. Ag/AgCl, a pulse duration of t_{pulse} = 3.0 s, a pause duration of t_{pause} = 0.1 s and a rectangular pulse shape [50]. Three different amounts of pulses were applied, namely 1, 8 and 16.

All involved physicochemical conditions are summarized in Figure 14 for better comparison.

$$\text{cathode:} \quad AuCl_4^- + 3e^- \rightarrow Au\downarrow + 4Cl^- \qquad E^\circ_{red} = 1.002\ V_{RHE} \tag{1}$$

$$\text{anode:} \quad 2Cl^- \qquad \rightarrow Cl_2 + 2e^- \qquad E^\circ_{ox} = 1.358\ V_{RHE} \tag{2}$$

$$6H_2O \qquad \rightarrow O_2 + 4H_3O^+ + 4e^- \quad E^\circ_{ox} = 1.229\ V_{RHE} \tag{3}$$

2.3. Characterization of Mesoporous TiO₂ Thin Films

Individual multi-layered MTTFs were characterized by means of UV/Vis and fluorescence spectroscopy as well as photoelectrochemical (PEC) water oxidation prior to a next deposition. Further experiments such as XRD, scanning electron microscopy (SEM) and nitrogen sorption were used for selected final samples only.

UV/Vis transmittance spectra in the range of λ = 200–800 nm were recorded with a *Varian Cary 4000* UV/Vis spectrophotometer. Prior to the actual measurement, baseline correction was ensured by measuring a cleaned blank FTO/glass substrate that has been calcined with the same temperature program as the investigated MTTF sample.

Fluorescence emissions were measured in the range of λ = 400–600 nm with a *Varian Cary Eclipse* fluorescence spectrophotometer. The excitation wavelength was 380 nm.

Photoelectrochemical (PEC) measurements were performed using an electrochemical workstation with a main potentiostat (*ZAHNER ZENNIUM Pro*) and a secondary potentiostat (*ZAHNER PP211*) to power a UV-LED (λ = 375 nm, $\delta\lambda_{FWHM}$ = ±7 nm, 70 W·m⁻² incident photon flux). The experiments were controlled with a computer and the *THALES XT* software package. A PEC cell (*ZAHNER PECC-2*) was filled with aqueous 1 M NaOH of pH = 13.6 and a three-electrode configuration was established by using a platinum wire, Ag/AgCl(sat. KCl) and the MTTF as counter, reference and working electrodes, respectively. Electrical contact to the films was made with help of copper wires and copper tapes attached to the bare FTO part of each sample. All experiments were performed under front-side illumination, thus the light entered the PEC cell from the electrolyte side through a quartz window. To eliminate interference by external irradiation, the PEC cell was placed into a light exclusion box.

For better comparison towards results from literature, the applied potentials versus Ag/AgCl(sat. KCl) were also translated into potentials relative to the reversible hydrogen electrode (RHE) by using the Nernst equation:

$$E_{\mathrm{RHE}} \ = \ E_{\mathrm{Ag/AgCl(sat.\ KCl)}} + 0.059\ \mathrm{V\ pH} \ + \ E^{\circ}_{\mathrm{Ag/AgCl(sat.\ KCl)}} \tag{4}$$

$$\mathrm{with}\ E^{\circ}_{\mathrm{Ag/AgCl(sat.\ KCl)}} \ = \ 0.1976\ \mathrm{V}$$

$$\mathrm{and\ pH} \ = \ 13.6$$

Chronoamperometry (*It*) under UV irradiation was performed with six illumination-dark-periods of 10 s each at a fixed potential of 1.20 V versus RHE. Linear voltammetric scans, or briefly (*IV*) scans, were carried out under UV irradiation or in the dark to ensure that no impurity-related adulterations of the *It* data was present. Accordingly, voltage scans were performed in the range of +0.4 V to +2.25 V versus RHE with a scan rate of 15 mV·s^{-1}. Prior to photoelectrochemical measurements of gold-loaded MTTFs, the electrolyte had to be purged for 30 min with nitrogen to replace dissolved air oxygen. This step is necessary to eliminate the reductive dark current related to the oxygen reduction reaction.

Cleaning was necessary to remove characterization-related impurities prior to the deposition of an additional MTTF layer. Conductive adhesive and remaining NaOH from the PEC characterization were removed by acetone and rinsing with de-ionized water, respectively. The cleaning was completed by a short period of ultrasonication in de-ionized water and an overnight drying at 100 °C.

Scanning electron microscopy (SEM) was carried out using a *FEI Helios NanoLab 600i*. Prior to the experiment, substrates were cut and affixed onto aluminum sample holders using conductive carbon tape and copper wires before applying conductive silver adhesive. SEM images of both surfaces and breaking edges were recorded with a secondary electron detector in a working distance of 4 mm and with an acceleration voltage of 10 kV and an electron beam current of 0.17 nA.

Combined energy dispersive X-ray spectroscopy and scanning electron microscopy (EDX-SEM) was used for gold-loaded MTTFs. Acceleration voltage and electron beam current were raised to 20 kV and 5.5 nA, respectively. F, Si, Ti, Sn and Au were selected as elements to assign and overlay pictures of the collected EDX mappings were generated with the *EDX Genesis* software.

Grazing incident X-ray diffractograms (GI-XRD) were obtained with a *PANalytical Empyrean* diffractometer with CuK$_\alpha$ = 1.54 Å radiation at an incident angle of 0.5°. The Scherrer equation was applied to determine the average crystallite sizes.

Nitrogen adsorption-desorption was measured with a *Micromeritics ASAP 2020* device. Prior to the adsorption measurements, the samples were degassed at 150 °C for 2 h at a base pressure of 2×10^{-6} mm Hg. For each measurement, two identically prepared MTTF substrates were placed inside a special substrate holder. Surface areas calculated according to the formalism established by Brunauer, Emmett and Teller (BET) from two selected partial pressures (0.005 p/p$_0$ and 0.038 p/p$_0$) and were related to the MTTFs' geometric surfaces due to the very low and in general hard to determine mass of individual films. The two-point calculation was selected in all cases since it was only possibly to collect full sorption isotherms for samples with >4 layers. Pore size distributions were calculated from the desorption branches of the isotherms using the Barrett–Joyner–Halenda (BJH) model.

3. Results and Discussion

3.1. Single-Layered Films

Figure 3 shows top-view SEM images of four single-layered MTTFs after different calcination temperatures between 400–550 °C. After the 400 °C treatment, the desired well-ordered mesoporous structure was obtained which is in good agreement with the protocol by Nagpure et al. [77]. With higher temperatures, the mesostructure becomes increasingly deteriorated, involving wall collapses and particle agglomeration. This observation that a formerly pristine network consisting of closely packed micelles within an amorphous TiO_2 matrix experiences structural deterioration (and formation of worm-like structures) upon air calcination has been commonly observed for MTTFs and is explained in terms of mass migration that accompanies the proceeding crystallization [62,80]. At the macroscopic scale, higher temperature treatment affected the structure of the film leading to cracks encompassing larger regions of FTO which appear much brighter than the TiO_2 in the SEM images. The measured thickness was around 450 nm for a single-layered film calcined at 550 °C (shown later in Figure 7). This is almost twice as much as the ≈250 nm mentioned by Rankin et al. [77] who reported the use of an identical EISA-sol composition as well as similar aging/thermal treatment. Unfortunately, these authors did not mention their dip-coating conditions (most importantly the used withdrawal rate). Other thicknesses were not determined within this study, but it was shown elsewhere that the thermally induced structural collapse is accompanied with a shrinkage of films [81].

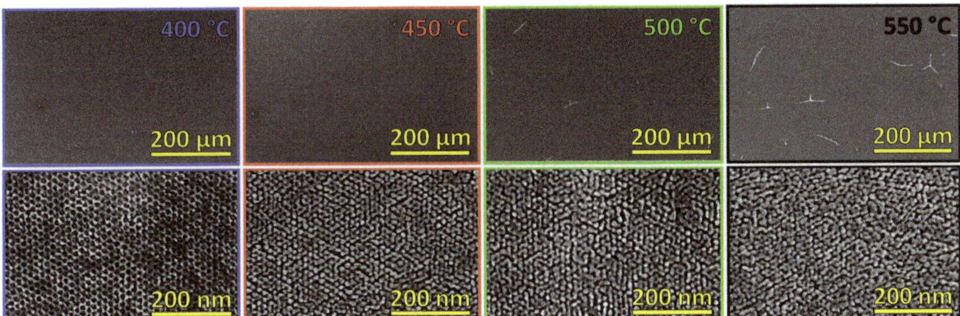

Figure 3. Influence of calcination temperature on mesofilm morphology as observed by top-view SEM. Higher calcination temperatures cause mesostructure deteriorations and the formation of cracks at T > 500 °C.

To analyze phase composition and crystallinity as a function of calcination temperature, grazing incidence X-ray diffraction was used. The obtained diffractograms for single-layered MTTFs treated at different temperatures between 350–550 °C are shown in Figure 4. Reference patterns of anatase TiO_2 [82], rutile TiO_2 [83] and SnO_2 [84] (as a representative for FTO) are included for reflex assignment and a diffractogram of a blank FTO/glass substrate was measured to identify the contribution of the FTO to the recorded diffractograms.

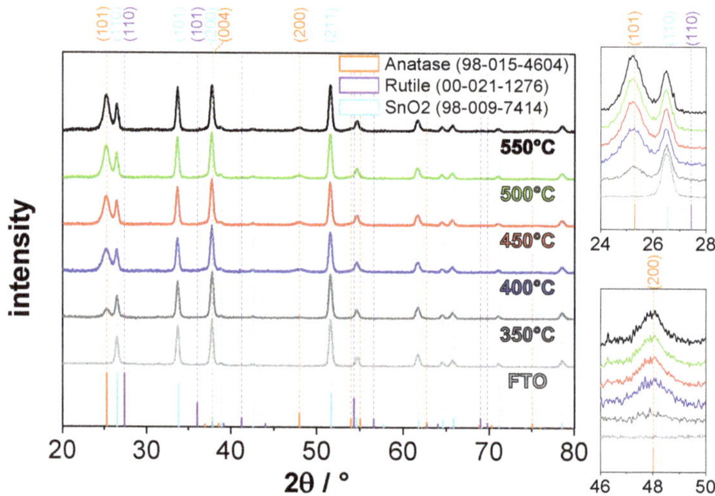

Figure 4. Normalized GI-XRD patterns of single-layered MTTFs calcined at different temperatures. Reference patterns [82–84] and selected hkl values are displayed for better interpretation.

The observed reflexes in the blank FTO diffractogram can be explained with the used reference pattern of SnO_2 in its rutile modification [84]. It is worth mentioning that the (200) orientation at 37.5° of 2θ is preferential in the used FTO/glass substrates. After a 350 °C thermal treatment, a weak anatase (101) reflex at 25.3° of 2θ is already visible but FTO-related reflexes dominate (e.g., those related to the (110), (101), (200) and (211) planes at 26.3°, 33.7°, 37.5° and 51.3° of 2θ, respectively). Those reflexes are part of all diffractograms of single-layered MTTFs as result of the small TiO_2 thickness. However, they start to diminish with additionally deposited layers (shown later in Figure 8). With higher calcination temperatures, the anatase phase becomes more crystalline as more intense (still broad) signals are observed (see insets). Average crystallite sizes for the most pronounced anatase reflex, namely (101), were calculated using the Scherrer equation and the results are summarized in Table 1. The observed average grain size increases for elevated calcination temperatures from 16 nm to 28 nm for samples treated at 400 and 550 °C, respectively are caused by mass migration and particle agglomeration during crystallization [62,80]. This agrees with the structural deterioration observed from the SEM imaging. No formation of rutile was observed in the investigated temperature range, indicating anatase is the exclusive phase in the MTTF. For instance, the most prominent rutile reflex, namely the (110) plane expected at 27.4° of 2θ, was not formed at all (see upper inset in Figure 4). However, also less pronounced rutile reflexes such as its (101), (111), and (301) plane (expected around round 36.0°, 41.2° and 69.0° of 2θ, respectively) were absent. This is in accordance with previous reports, where highly thermally stable MTTFs were investigated and the formation of rutile was not observed for temperatures below 700 °C [60,85,86].

Table 1. Analysis of anatase (101) GI-XRD peak broadening of single-layered MTTFs calcined at different temperatures.

T/°C	FWHM/°	Average Crystallite Size/nm
400	0.859	≈16
500	0.709	≈24
550	0.638	≈28

The photoactivity of single-layered films was measured by means of photoelectrochemical water oxidation under UV illumination. Linear voltammetric scans of thermally treated MTTFs at different temperatures are shown in Figure 5.

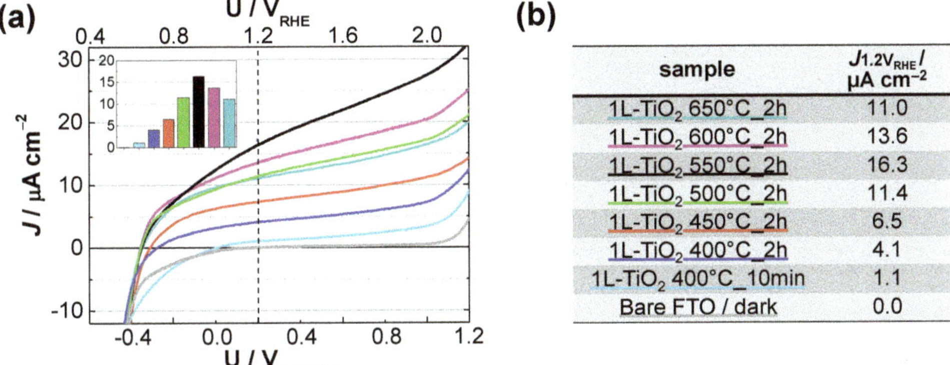

sample	J1.2V$_{RHE}$ / µA cm^{-2}
1L-TiO$_2$ 650°C_2h	11.0
1L-TiO$_2$ 600°C_2h	13.6
1L-TiO$_2$ 550°C_2h	16.3
1L-TiO$_2$ 500°C_2h	11.4
1L-TiO$_2$ 450°C_2h	6.5
1L-TiO$_2$ 400°C_2h	4.1
1L-TiO$_2$ 400°C_10min	1.1
Bare FTO / dark	0.0

Figure 5. Influence of calcination temperature on photoelectrochemical water oxidation activity. (**a**) Voltammetric scans and (**b**) selected photocurrent densities for each corresponding calcination program.

The observed photocurrent densities strongly depend on the chosen calcination program and reach a maximum for the 550 °C treated samples. Higher calcination temperatures result in decreased photocurrent densities, which is likely related to a larger degree of mesostructural collapse that can no longer be overcompensated by simultaneously improving crystallinity of the MTTF. Such trade-offs between improved optoelectrochemical properties on the one hand and loss of accessible active sites on the other are commonly observed for MTTFs [58,62]. The optimal temperature for a given photocatalytic device not only depends on the involved chemicals and their procession but also on its later purpose. In the context of this study, the calcination program with a holding temperature of 550 °C was chosen for the preparation of multi-layered films. This thermal treatment gives highest photoactivities due to improved crystallinity and a still accessible (albeit deteriorated) pore structure.

3.2. Multi-Layered Films

Multi-layered films with up to 10 layers have been prepared in terms of an "interrupted" layer-by-layer deposition which is schematically depicted in Figure 1 on page 3. Accordingly, a typical deposition cycle consisted of the dip-coating deposition, followed by a high humidity aging treatment at low temperatures, an overnight preconsolidation at 100 °C and finally the calcination at the optimized temperature of 550 °C. It is emphasized that the term "interrupted" derives from the fact that individual films were characterized prior to the next layer deposition. Such a characterization usually involved UV/Vis and fluorescence spectroscopy and—more importantly—photoelectrochemical characterization in terms of water oxidation in a highly alkaline media. This interim characterization/cleaning is meant to ensure (i) the sole effect of additional layers by avoiding sample variations and (ii) the robustness of the films without losing activities. However, its potential effect on the observed photoelectrocatalytical and structural properties must be considered in context of our experimental findings.

Figure 6 illustrates the optical properties of the obtained MTTFs. Part (a) shows a side-by-side photograph of selected multi-layered substrates. Increasing opacities in the visible range already indicate additional amounts of TiO$_2$ with each new layer.

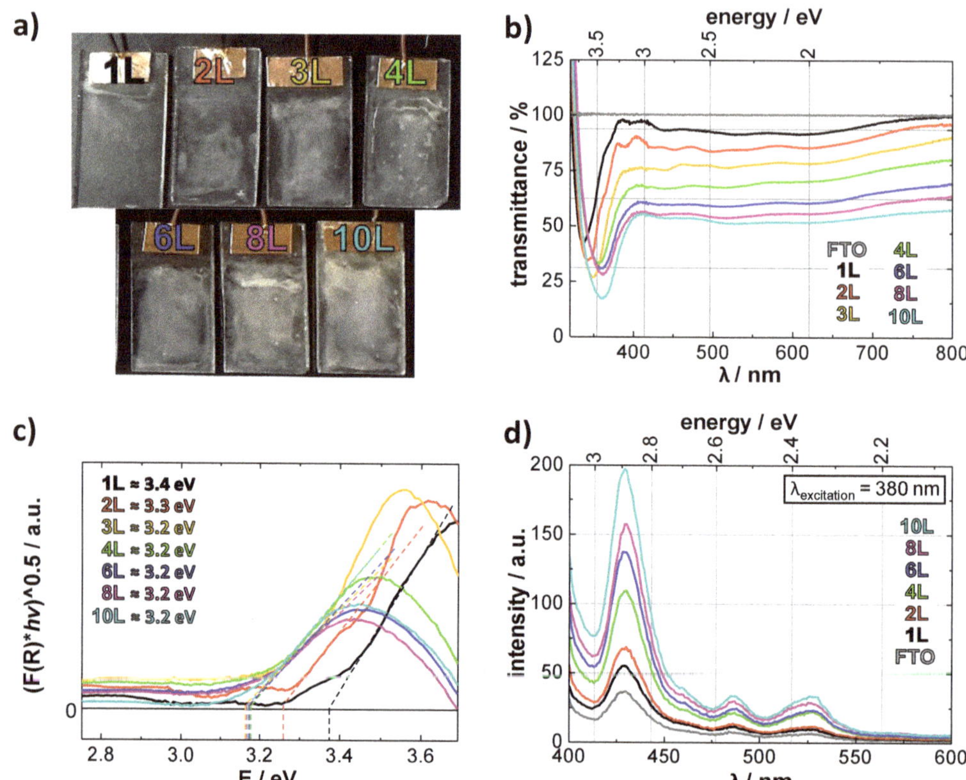

Figure 6. Photograph and optical spectra of multi-layered MTTFs calcined at 550 °C. (**a**) Photographs of selected samples, (**b**) UV/Vis transmittance spectra, (**c**) Tauc plots, and (**d**) fluorescence spectra.

This observation is confirmed in terms of the UV/Vis transmittance spectra shown in Figure 6b where the increasing opacity is apparent in both the UV and visible range. While reduced UV transmittance is entirely intentional and the decisive motivation for multi-layer formation in the first place, the optical opacity might be beneficial for intended future incorporation of plasmonic nanoparticles. The intense back-scattering of visible photons within the film implies that more visible photons could be consumed for surface plasmon resonance.

Figure 6c shows the Tauc plots and optical band gaps. While single- and double-layered films show slightly higher band gaps of ≈3.4 eV and ≈3.3 eV respectively, films consisting of four and more layers display values close to the expected 3.2 eV of TiO_2 in its anatase modification [2]. This observation of slightly larger optical band gaps for 1L and 2L could hint at quantum size effects that disappear with particle growth due to the additional thermal processing of multi-layered films. Analysis of peak broadening from the Gi-XRD pattern (Figure 8) however, does not reveal increased grain sizes. In fact, the average crystallite sizes for the most dominant (101) reflexes remain relatively unaltered at 27 nm. However, these grains might consist of coalesced (aggregated) small nanoparticles, which only sinter completely during additional thermal treatment.

The deposition of additional TiO_2 is furthermore confirmed by fluorescence spectroscopy as increasingly intense fluorescence emissions indicate in Figure 6d. These results are only considered to be qualitative indicators for the presence of higher TiO_2 amounts and any further interpretation is not considered here. Briefly, the experience that one and the same sample occasionally shows noticeable fluctuations of fluorescence intensities gives reason for this perception.

A more reliable evidence for the additional deposition of TiO_2 is the development of film thickness which was determined by SEM imaging of breaking edges and is shown in Figure 7.

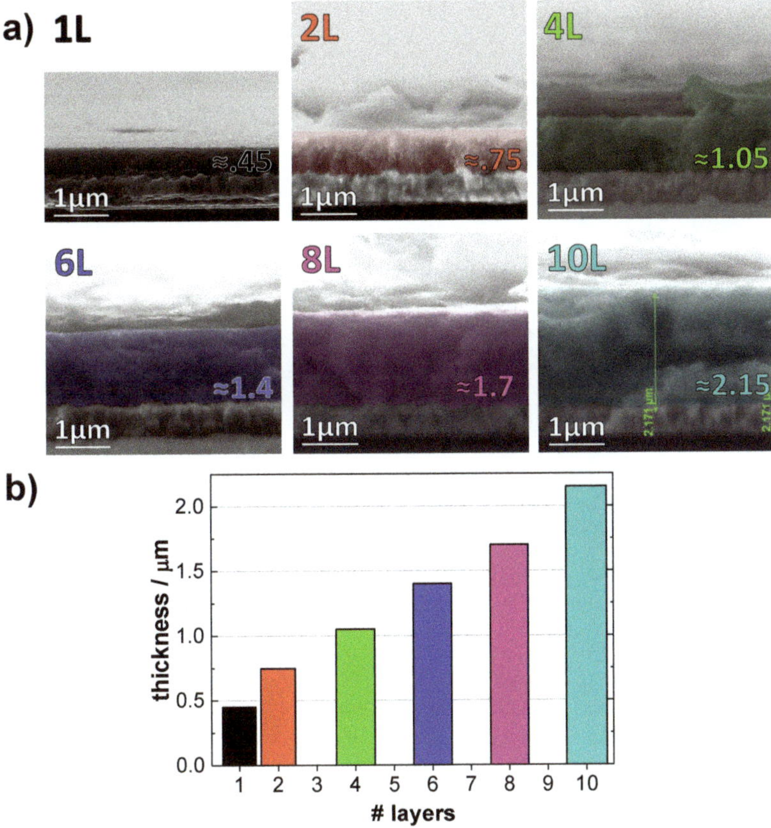

Figure 7. Development of film thickness as function of deposited MTTF layers. (**a**) Breaking edge SEM images of multi-layered MTTFs calcined at 550 °C and (**b**) their corresponding thickness-versus-layer plot.

The development of 0.45, 0.75, 1.05, 1.4, 1.7 and 2.15 µm for 1-, 2-, 4-, 6-, 8- and 10-layered films, respectively is in fact not as clearly indicative for a linear correlation between film thickness and number of deposited layers as claimed by other authors [47,73–75,87,88]; however, a subsequent gain of TiO_2 is evident. Considering the development from 2–10 layers, a linear increase (of ≈150–200 nm per layer) can be determined. Thus, only the first two layers with their above-the-average thickness exhibit an exception.

One possible explanation for the exceptionally high thickness of 0.45 µm of the first layer could be that the surface initially deposited onto (FTO) displays a significantly different environment compared to that of all subsequent layers (mesoporous TiO_2). Another explanation regarding the second layer could be related to the 550 °C air calcination and total oxidative removal of surfactants that fully exposes the porous structure of the first layer. This in turn might allow another EISA-sol infiltration compared to higher-layered MTTFs later. In fact, the Grosso group reported a layer-by-layer synthesis route involving an interposed dip-coating into a diluted ethanolic surfactant solution (just these two chemicals) to allow pore refilling of thermally exposed mesopores prior to subsequent dip-coating into EISA-sol [73]. Such a treatment allowed a linear increase of film thickness. For more than two layers indeed, a more or less linear dependence is observed indicating an "equilibrated sub-surface" for the subsequent deposition cycle.

Figure 8 shows GI-XRD patterns of the obtained multi-layered MTTFs together with reference patterns.

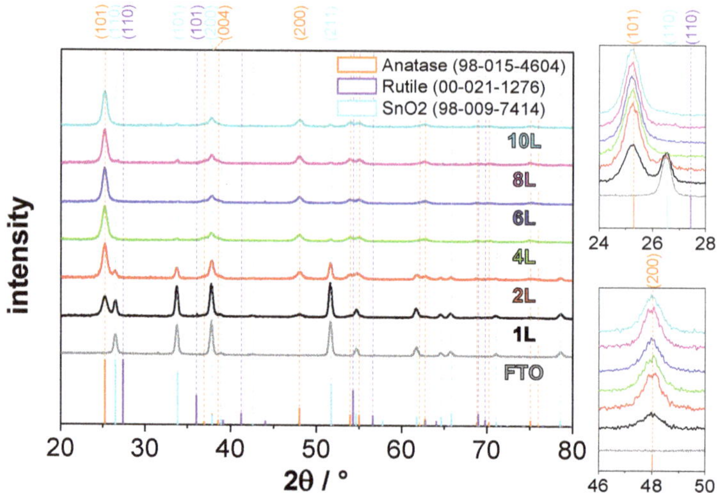

Figure 8. Normalized GI-XRD patterns of multi-layered MTTFs calcined at 550 °C. Reference patterns [82–84] and selected hkl values are displayed for better interpretation.

In case of single- and double-layered films, FTO-related reflexes are dominant, but they quickly decline with further layers. Still noticeable FTO reflexes such as (101) and (211) (at 33.7° and 51.3° of 2θ, respectively) for 10-layered films could possibly be related to the porous and non-compact TiO_2 structure. All multi-layered MTTFs appear to consist entirely of the anatase phase. An improved crystallinity in terms of sharper anatase reflexes as a result of the multiple thermal treatments was expected, since improved crystallinity of anatase within the bottom layers (resulting from repeated thermal processing with the additional layer) had already been demonstrated for a relatively lower interim temperature treatment at 350 °C [74]. The authors performed incident angle dependent GI-XRD scans between 0.1–0.4° on a five-layered MTTF, thus manipulating the X-ray's penetration depth and obtain insights on the crystallinity changes at the bottom layers. Similarly, the MTTFs investigated in our work may have improved bottom-layer crystallinity as the intermediate thermal treatment was higher (550 °C) than in the previous report [74]. The average crystallite sizes obtained from the most dominant (101) reflexes remain relatively unaltered at 27 nm (see Table 2).

Table 2. Analysis of anatase (101) GI-XRD peak broadening of single-layered MTTFs calcined at different temperatures.

	FWHM/°	Average Crystallite Size/nm
1 L	0.638	≈28
2 L	0.672	≈25
4 L	0.729	≈22
6 L	0.623	≈30
8 L	0.626	≈29
10 L	0.658	≈26

The specific surface area of the films was investigated with nitrogen sorption experiments and BET analysis. The BET surface areas were derived from two selected partial pressures (0.005 p/p_0 and 0.038 p/p_0) and were related to the MTTFs' geometric surfaces due to the very low and in general difficult to determine mass of individual films. Figure 9a shows the calculated BET surfaces as a function of the number of deposited MTTF layers.

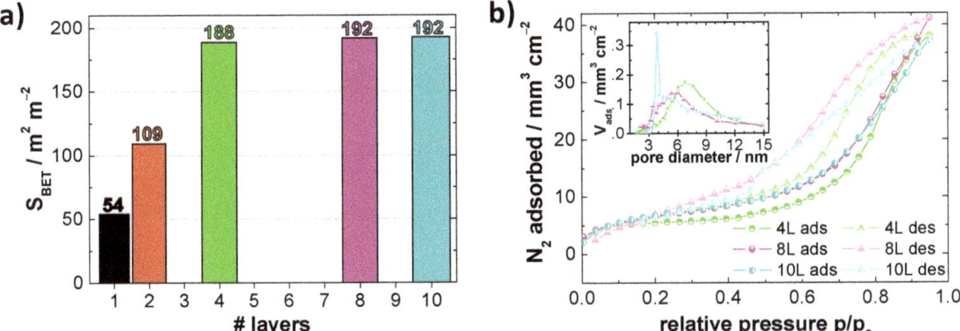

Figure 9. Nitrogen adsorption-desorption isotherms of multi-layered MTTFs calcined at 550 °C. (**a**) The calculated BET surface areas as a function of the number of deposited MTTF layers and (**b**) nitrogen sorption isotherms and pore size distribution of selected substrates.

Surface areas increased almost linearly up to four layers where it reached 188 $m^2 \cdot m^{-2}$. Beyond that almost no changes were observed. However, the trend of quickly saturated surface areas is in very good agreement with similar observation by Procházka et al. [74]. They studied P123-templated MTTFs with 1–5 layers which were obtained via layer-by-layer dip-coating onto FTO/glass and with 2 h lasting thermal treatment at 350 °C prior to the next deposition cycle (but without interruption by characterization/cleaning). Their finding of an unchanged surface area for more than 3 layers was explained by the compensation of two opposing processes: added surface area per deposition and lost surface area due sintering of bottom layers upon thermal treatment [74]. This very explanation is likely to apply here as well. Moreover, they documented two additional features: the stepwise improvement of the crystallinity of anatase nanoparticles within bottom layers during repeated thermal processing and the formation of denser crust films surfaces as a result of one-dimensional constrained sintering. Although these effects might apply here as well, they are not clearly pronounced in the collected XRD and SEM data, respectively as stated before. Figure 9b presents full nitrogen sorption isotherms of selected multi-layer films (4 L, 8 L, 10 L) showing mesopore condensation. An inset shows pore size distribution from desorption isotherms. With additional layers (and additional thermal treatments), the pore size distributions shift to smaller values indicating pore narrowing due to "wall thickening" during the subsequent sol infiltration process.

To correlate the structural properties of MTTFs involved in the present study with their activity, PEC water oxidation in a highly alkaline media was used as a test reaction. More precisely, chronoamperometric experiments were carried out under controlled chopped-light UV illumination conditions at 1.2 V versus RHE. For a better reliability, two identically prepared samples were tested. The corresponding experimental results are depicted in Figure 10.

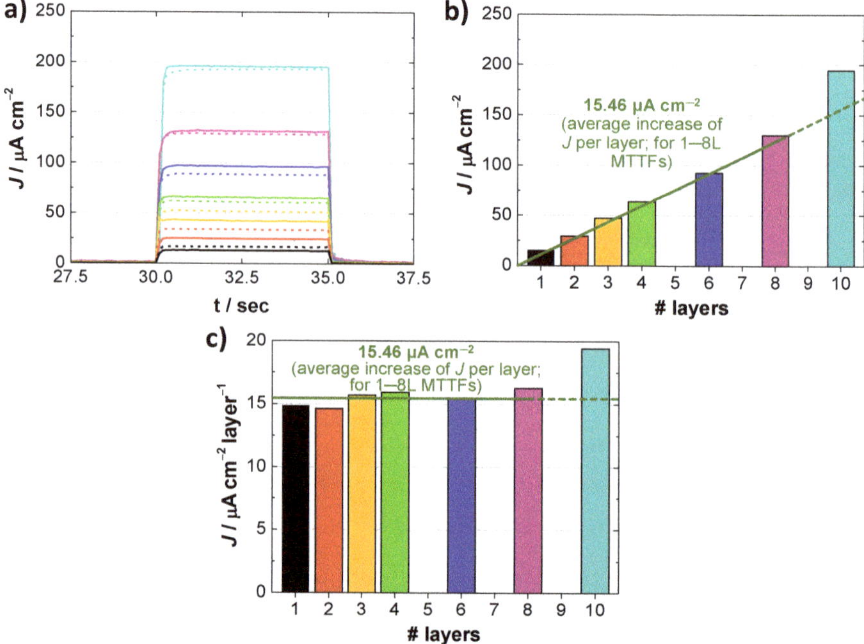

Figure 10. Photocurrent densities (during PEC water oxidation) of multi-layered MTTFs calcined at 550 °C as a function of the number of deposited MTTF layers. (**a**) The corresponding chronoamperometric measurements. (**b**) Average values of the two sets of samples used during the chronoamperometric experiment. A green line indicates the average activity increase for the first eight layers. (**c**) derivative representation of (**b**) to emphasize the specialty of both 10L films.

From the raw data depicted in Figure 10a it is clearly visible that the photocurrent increases with the number of layers. The similarity of results for a pair of samples with the same number of layers (denoted as dashed and solid lines of same color) underlines their good reproducibility. Figure 10b shows the average photocurrents of each pair as a function of the number of layers. It becomes clearly visible that the activity increases almost linearly in the range of 1–8 layers, leading to an average increase of $\approx 15.5\ \mu A \cdot cm^{-2}$ per layer. To our surprise, the 10-layered MTTFs exceed this trend by exhibiting a significantly higher photocurrent per layer value of $\approx 19.4\ \mu A \cdot cm^{-2}\ layer^{-1}$. Figure 10c further emphasizes this circumstance in terms of a derivate plot.

Without focusing too much on the exceptionally active 10-layered films, the observation of an undiminished activity increases up to 10 L (if even beyond was not tested) has to be explained in the context of the already discussed findings: One is that the thickness increases almost linearly (at least if single- and double-layered films are not considered). Another is that the surface area is almost constant (again if single- and double-layered MTTFs are not considered). Therefore, the enhanced activity needs to be explained by access to more active sites derived from improved optoelectrochemical properties and decreased recombinational losses in bottom layers near the FTO. Additionally, the higher rates of charge carrier formation due to stronger absorption of thicker materials contribute as well. It should be pointed out that the same thickness-versus-surface-area behavior was observed and successfully explained in terms of increased bottom-layer crystallinity before [74].

A further possible explanation could be related to the "interrupted" layer-by-layer deposition. In fact, it may induce structural altering as result of characterization/cleaning steps between each deposition. Figure 11 shows top-view SEM images of single- versus ten-layered MTTFs at four different magnifications.

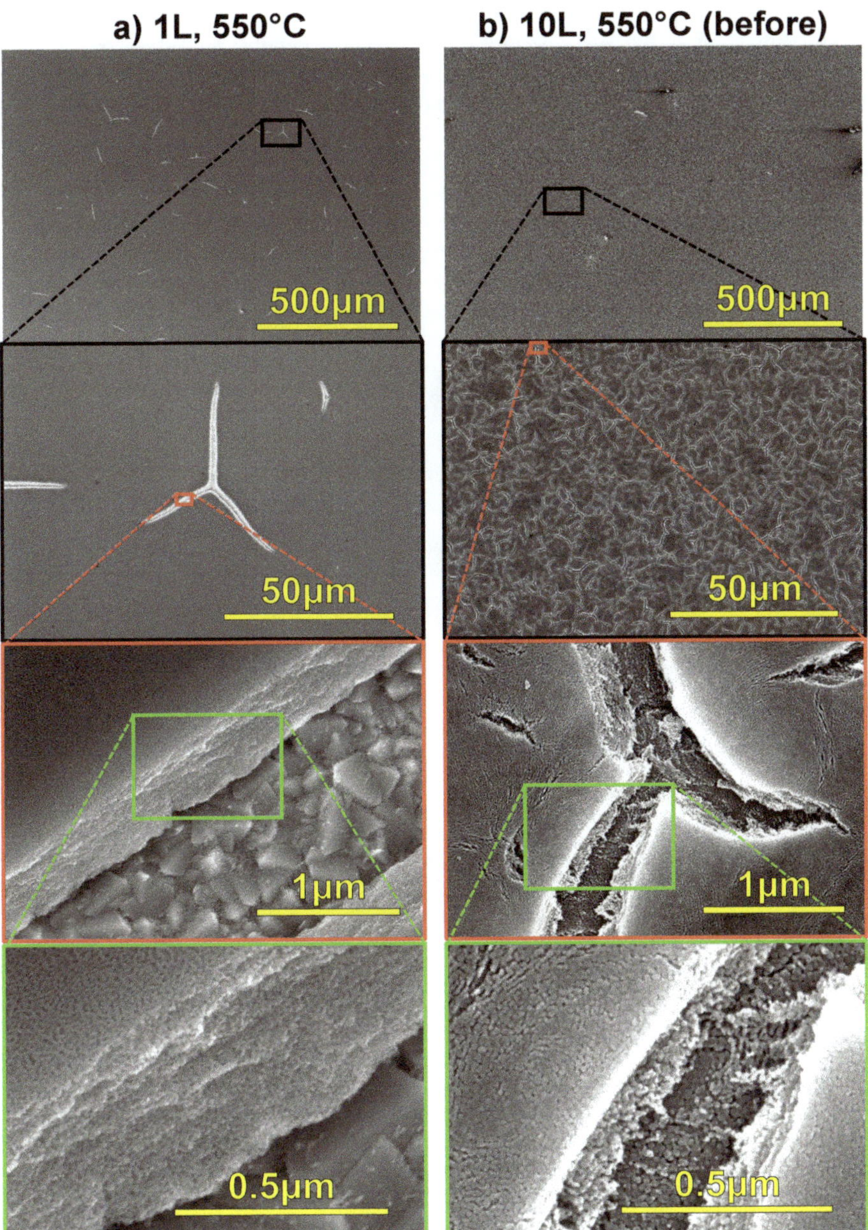

Figure 11. SEM top-view image of (**a**) single- and (**b**) ten-layered MTTFs at different magnifications showing that an extended network of microscopic cracks is formed in case of the multi-layered sample.

Single-layered films show some isolated cracks which expose the FTO back contact (white lines in the images). The topology of the ten-layered films, however, was not expected. SEM imaging reveals the formation of an extended network of small cracks which are much smaller in size but much more prominent than those for single-layered films. Topologically, these structures resemble a hierarchically macro-mesoporous systems that were prepared by controlled phase separation and surfactant templating [89]. Such a structure provides improved transport properties as the electrolyte's infiltration/diffusion into the MTTF becomes facilitated and therefore is likely to contribute to the enhanced activity as well. At this point it must be determined in terms of uninterrupted control experiments whether this extensive crack formation is related to the characterization/cleaning step between deposition or if it is caused by the multiple calcination steps.

3.3. Multi-Layered Films Stability

To gain insights on the chemical and mechanical stability of the films, we have performed different sets of stability measurements. Electrochemical stability tests were performed on 10-layered MTTFs and involved the steps of disassembling, cleaning, drying and assembling between each chronoamperometry measurement. These measurements were carried out for ten consecutive days and—for better reproducibility—two sets of samples treated at 450 °C and 550 °C (two films each) were used (see Figure 12a). At the end of the tenth measurement, the sample "550 °C_A" was mechanically rubbed with a cotton bud (Q-Tip$^{©}$) (see Figure 13a).

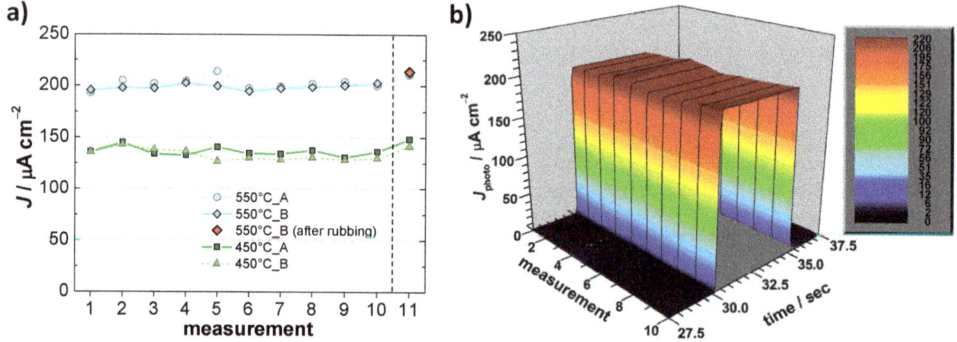

Figure 12. (**a**) Photocurrent stability tests of four 10-layered MTTFs (2× 450 °C, 2× 550 °C) recorded at ten different days (disassembling, cleaning, drying, assembling between each measurement). Photocurrent densities were derived from chronoamperometric experiments and taken after 32.5 s (right in the middle of an illumination phase). (**b**) Three-dimensional representation of the chronoamperometric results that correspond to a the "550 °C_A" sample from (**a**).

As shown in Figure 12a,b, all films deliver stable photocurrents during the first 10 experiments. Films treated at 450 °C show values around ≈140 µA·cm^{-2} and films treated at 550 °C ≈ 200 µA·cm^{-2}. To set this into perspective, stable photocatalytic activities of MTTFs (loaded with Pt NPs) were reported previously by Ismail for both long term and repeated measurements of the photocatalytic gas-phase oxidation of acetaldehyde [75].

After the rubbing test of "550 °C_A" (red diamond), still comparable photocurrents are observed with very slight variations, suggesting the films are quite robust and intact. The slight increase of photocurrents at measurement 11 might be related to the use of a freshly prepared new electrolyte solution. When considering the photographs taken before and after rubbing, the applied mechanical stress did not result in macroscopic changes of the films (see Figure 13a). Corresponding SEM micrographs, however, show that the structural integrity of the films is affected to an extent as scratch marks and microstructural changes reveal. Overall, there seems to be no significant material abrasion since the activity stays nearly unaffected which confirms the good adhesion. However, MTTFs cannot withstand a scotch tape adhesion test (not shown).

In hope for an accelerated film preparation process, two sets of comparative experiments were carried out: In the first set, we have prepared multi-layered films without an intermediate temperature step and tested the mechanical stability in a similar way. Unfortunately, these films are not mechanically stable at all and can be peeled off easily (see Figure 13b). It is worth mentioning that a very comparable synthesis route (also without intermediate calcination) was recently reported by Rankin et al. They used MTTFs with up to five layers as negative electrodes in Li-ion batteries and did not report stability problems [88]. In another attempt towards faster production of thicker films, a more viscous EISA-sol was used by having a lower amount of the solvent EtOH in the initial EISA-sol. The obtained MTTFs exhibited the same problem of bad adhesion (photograph not shown).

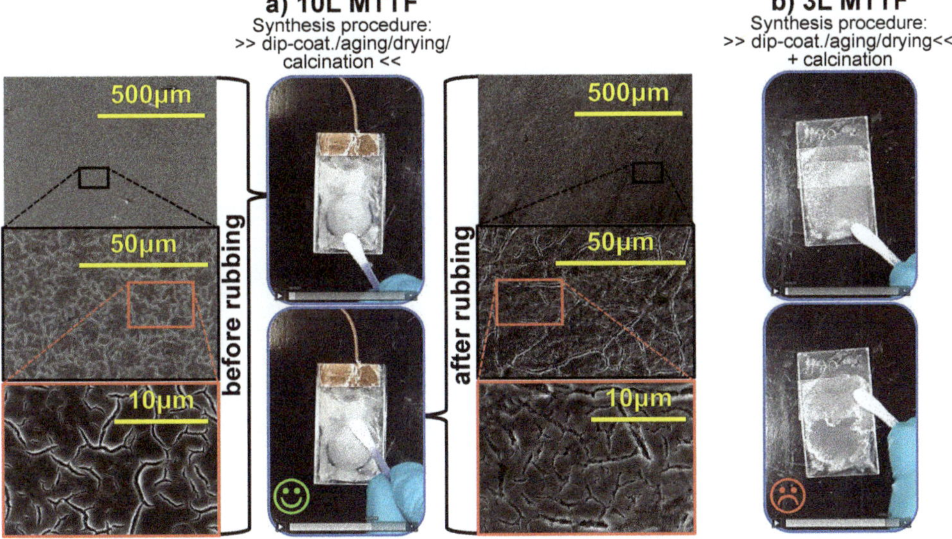

Figure 13. Two different samples before and after the "rubbing test". (**a**) Photograph and SEM top-view images of the 10-layered MTTF "550 °C_A" from Figure 12. Although its microscopic structure is affected by the rubbing, this sample has delivered stable photocurrents prior and after this procedure. (**b**) Photograph of a 3-layered MTTF that had been calcined only once to save energy and time, namely after the drying phase of its third and last deposition cycle. This film is not mechanically stable.

3.4. Proof of Concept: SPR-Active Gold Nanoparticles Inside MTTFs

In this brief subsection, the suitability of our MTTFs as porous host systems for the incorporation of SPR-active noble metal particles is investigated. As a test reaction, the cathodic pulsed electrodeposition (pulsed-ED) of gold nanoparticles was used.

The pulsed-ED technique provides the benefit of being simple to operate but still versatile, since it allows introduction of Au nanoparticles of different size/shape by use of different deposition potentials, pulse/pause sequences, pulse numbers etc. [50,90]. Therefore, pulsed-ED enables both fine-tuning of plasmonic properties and intense contact areas between TiO_2 and Au the same time. Individual single-layered MTTFs were treated with three different amounts of pulses, namely 1, 8 and 16.

Figure 14 shows the results of the experiments. Figure 14a summarizes the deposition bath composition and electrochemical pulse parameters. Optical analysis is shown in Figure 14b, where both a photograph and the corresponding UV/Vis transmittance spectra show the surface plasmon resonance features related to the introduced gold particles. Specifically, an SPR band is present for all three Au-modified samples and is in the green region of the visible electromagnetic spectrum around wavelengths of 530 nm. The tendency towards slightly increased wavelengths after more applied pulses goes along with the expected particle growth and the size (and shape) dependence of SPR features [57]. Furthermore, it should be pointed out that the porous TiO_2 host structure confines the growth of incorporated gold inside the film—with the limit being dendritic Au structures as partial replica of the original pore system [50].

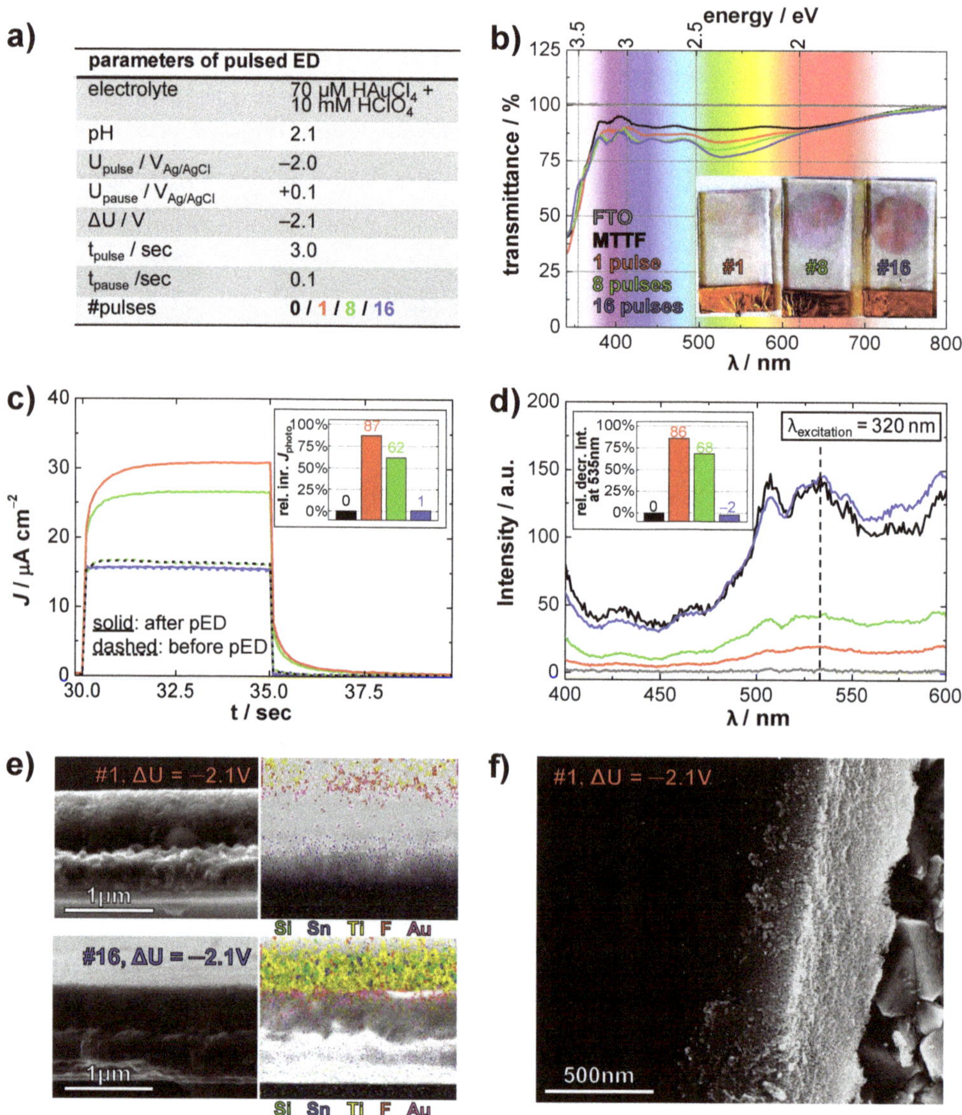

Figure 14. Properties of SPR-active gold-modified MTTFs obtained after three different amounts of applied cathodic pulses. (**a**) physicochemical conditions and pulse parameters during pulsed electrodeposition, (**b**) UV/Vis transmittance spectra and photograph, (**c**) chronoamperometric results (dashed and solid lines before and after pulsed-ED, respectively) with inset showing relative increase of photocurrents, (**d**) fluorescence spectra with inset showing relative decrease of emission intensity, (**e**) cross-sectional SEM images and corresponding EDX mapping of two selected samples, (**f**) top-view SEM image of most active sample near a cracked part showing Au-NPs.

The activity before and after the pulsed-ED was determined in terms PEC water oxidation under UV illumination and the corresponding chronoamperometric measurements are shown Figure 14c. The data show that exposure to 1, 8 and 16 pulses leads to highest, second highest and lowest photocurrents, respectively with corresponding relative photocurrent increases of 87%, 62% and 1%

(inset of Figure 14c). These increased activities are explained by the formation of a Schottky barrier at the Au–TiO$_2$ interface which causes spatial separation of photogenerated charge carriers involving transfer of electrons towards the noble metal [41,49]. The consequence is a reduced rate of charge carrier recombination which extends the amount of harvested charge carriers and their participation in photoelectrochemical reaction—ultimately leading to the observed improved photocurrents. This is astonishingly well confirmed in terms fluorescence emissions whose relative decreases of 86%, 68% and –2% (shown in Figure 14d) go almost perfectly hand in hand with the relative photocurrent increases. Both experiments prove that the formation of intense contact areas between the noble metal and the semiconductor is possible with the chosen pulsed-ED approach. The observed decreased activity for additional pulses might be a result of pore blocking or screening due to an excessive amount of gold. In fact, MTTFs that experienced an extended electrochemical treatment involving $\gg$16 pulses have an extensive deposition of Au on the film surface (not shown).

To further demonstrate the successful incorporation and to possibly obtain first insights regarding spatial distributions of Au-NPs, SEM-EDX mapping was performed on breaking edges of the samples exposed to 1 and 16 pulses. The corresponding SEM images and SEM-EDX overlays are shown in Figure 14e. Regions related to glass, FTO and TiO$_2$ can be easily identified and distinguished. More importantly, the incorporation of Au-NPs is indicated but information related to their spatial distribution (and their structural properties including size distribution) cannot be extracted from those images. It appears from the presented overlays that Au is enriched at the film surface. SEM-EDX mapping confirms the incorporation of Au but for more detailed information of spatial and size distribution, selected thin Au/MTTF filaments should be analyzed by transmission electron microscopy (TEM) which is intended for future investigation. Finally, a top-view SEM image of the most active sample is shown in Figure 14f. It was measured nearby a crack in the MTTF and the incorporation of Au-NPs is confirmed once more as they appear as small dots with much brighter contrast compared to the TiO$_2$ or the FTO.

4. Conclusions

Multi-layered mesoporous TiO$_2$ thin film (photo)electrodes with up to 10 layers were prepared by an EISA method and repeated dip-coating/aging/drying/calcination cycles ("interrupted" layer-by-layer deposition). The experimental conditions were derived from an existing route towards single-layered films and were optimized with respect to the calcination temperature.

Each additional layer was accompanied with an almost linear increase of activity due to increased amounts of immobilized TiO$_2$ and access to more active sites as a combined effect of increased surface area and better crystallinity. While additional TiO$_2$ causes stronger absorption of UV photons and thus more photogenerated e$^-$–h$^+$ pairs, accessible porous structure plus improved crystallinity allows their collection and participation in photochemical reactions. Another contribution is related to improved transport properties due to an extended network of cracks which likely derived from the interim characterization/cleaning steps. These combined effects explain the enhanced activity. Since 10-layered films show the highest photocurrents in this study, there is no indication of an already saturated activity which implies that the it can be further maximized with additional layers.

Moreover, the 10-layered films possess both a good photoelectrocatalytical and mechanical stability making them suitable candidates for future modifications. It was further shown that two faster routes towards thicker films lead to poor adhesion (an easy peeling off) rendering them useless for a further usage.

Initial results show the suitability of MTTFs as porous host substrates. Plasmonic gold nanoparticles were incorporated via pulsed electrochemical deposition and the presence of Au-NPs was confirmed (SEM-EDX, UV/Vis spectroscopy). All nanocomposites showed improved activities under UV illumination compared to pure MTTFs which is in excellent agreement with decreased fluorescence emissions. These observations are explained by reduced h$^+$–e$^-$ rates the Au–TiO$_2$

interface (Schottky barrier). Further study is in progress to get in depth insights at structural properties of noble metal deposits, visible-light responses and optimum number of layers of the MTTF host.

Author Contributions: Conceptualization, E.G., D.H.T. and M.W.; Methodology, E.G. and D.H.T.; Software, E.G.; Validation, E.G. and D.H.T.; Formal Analysis, E.G. and D.H.T.; Investigation, E.G.; Resources, M.W.; Data Curation, E.G. and D.H.T.; Writing—Original Draft Preparation, E.G.; Writing—Review and Editing, E.G. and D.H.T.; Visualization, E.G.; Supervision, M.W.; Project Administration, M.W.; Funding Acquisition, M.W.

Funding: This research was funded by the Deutsche Forschungsgemeinschaft (DFG) within the SPP 1613 (WA 1116/28-1) and INST 184/154-1 FUGG and the Bundesministerium für Bildung und Forschung (BMBF) within the project *PROPHECY* (*PRO*zesskonzepte für die *PH*otokatalytische CO_2-Reduktion verbunden mit *LifE-CY*cle-Analysis).

Acknowledgments: The authors further thank Lea Mohrmann who was significantly involved in the preparation of multi-layered MTTFs.

Conflicts of Interest: The authors declare no conflict of interest.

References

1. Fujishima, A.; Honda, K. Electrochemical Photolysis of Water at a Semiconductor Electrode. *Nature* **1972**, *238*, 37–38. [CrossRef] [PubMed]
2. Diebold, U. The surface science of titanium dioxide. *Surf. Sci. Rep.* **2003**, *48*, 53–229. [CrossRef]
3. Fujishima, A.; Zhang, X.; Tryk, D.A. TiO_2 photocatalysis and related surface phenomena. *Surf. Sci. Rep.* **2008**, *63*, 515–582. [CrossRef]
4. Fattakhova-Rohlfing, D.; Zaleska, A.; Bein, T. Three-Dimensional Titanium Dioxide Nanomaterials. *Chem. Rev.* **2014**, *114*, 9487–9558. [CrossRef]
5. Fujishima, A.; Rao, T.N.; Tryk, D.A. Titanium dioxide photocatalysis. *J. Photochem. Photobiol. C* **2000**, *1*, 1–21. [CrossRef]
6. Hashimoto, K.; Irie, H.; Fujishima, A. TiO_2 Photocatalysis: A Historical Overview and Future Prospects. *Jpn. J. Appl. Phys.* **2005**, *44*, 8269–8285. [CrossRef]
7. Haider, A.J.; Jameel, Z.N.; Al-Hussaini, I.H. Review on: Titanium Dioxide Applications. *Energy Procedia* **2019**, *157*, 17–29. [CrossRef]
8. Bai, J.; Zhou, B. Titanium Dioxide Nanomaterials for Sensor Applications. *Chem. Rev.* **2014**, *114*, 10131–10176. [CrossRef]
9. Wang, Y.; Wu, T.; Zhou, Y.; Meng, C.; Zhu, W.; Liu, L. TiO_2-Based Nanoheterostructures for Promoting Gas Sensitivity Performance: Designs, Developments, and Prospects. *Sensors* **2017**, *17*, 1971. [CrossRef]
10. O'Regan, B.; Grätzel, M. A low-cost, high-efficiency solar cell based on dye-sensitized colloidal TiO_2 films. *Nature* **1991**, *353*, 737–740. [CrossRef]
11. Gong, J.; Sumathy, K.; Qiao, Q.; Zhou, Z. Review on dye-sensitized solar cells (DSSCs): Advanced techniques and research trends. *Renew. Sustain. Energ. Rev.* **2017**, *68*, 234–246. [CrossRef]
12. Liu, Y.; Yang, Y. Recent Progress of TiO_2-Based Anodes for Li Ion Batteries. *J. Nanomat.* **2016**, *2016*, 8123652. [CrossRef]
13. Madian, M.; Eychmüller, A.; Giebeler, L. Current Advances in TiO_2-Based Nanostructure Electrodes for High Performance Lithium Ion Batteries. *Batteries* **2018**, *4*, 7. [CrossRef]
14. Nakajima, A.; Hashimoto, K.; Watanabe, T.; Takai, K.; Yamauchi, G.; Fujishima, A. Transparent Superhydrophobic Thin Films with Self-Cleaning Properties. *Langmuir* **2000**, *16*, 7044–7047. [CrossRef]
15. Ollis, D.F. Photocatalytic purification and remediation of contaminated air and water. *CR ACAD Sci. IIC Chem.* **2000**, *3*, 405–411. [CrossRef]
16. Hay, S.O.; Obee, T.; Luo, Z.; Jiang, T.; Meng, Y.; He, J.; Murphy, S.C.; Suib, S. The Viability of Photocatalysis for Air Purification. *Molecules* **2015**, *20*, 1319–1356. [CrossRef] [PubMed]
17. Yuan, J.; Fujisawa, R.; Tsujikawa, S. Photopotentials of Copper Coated with TiO_2 by Sol-Gel Method. *Zairyo-to-Kankyo* **1994**, *43*, 433–440. [CrossRef]
18. Sahnesarayi, M.; Sarpoolaky, H.; Rastegari, S. Influence of Multiple Coating and Heat Treatment Cycles on the Performance of a Nano-TiO_2 Coating in the Protection of 316L Stainless Steel Against Corrosion under UV Illumination and Dark Conditions. *Iran. J. Mater. Sci. Eng.* **2019**, *16*, 33–42. [CrossRef]

19. Dhakshinamoorthy, A.; Navalon, S.; Corma, A.; Garcia, H. Photocatalytic CO_2 reduction by TiO_2 and related titanium containing solids. *Energy Environ. Sci.* **2012**, *5*, 9217–9233. [CrossRef]

20. Wei, L.; Yu, C.; Zhang, Q.; Liu, H.; Wang, Y. TiO_2-based heterojunction photocatalysts for photocatalytic reduction of CO_2 into solar fuels. *J. Mater. Chem. A* **2018**, *6*, 22411–22436. [CrossRef]

21. Li, X.; Yu, J.; Jaroniec, M.; Chen, X. Cocatalysts for Selective Photoreduction of CO_2 into Solar Fuels. *Chem. Rev.* **2019**, *119*, 3962–4179. [CrossRef] [PubMed]

22. Osterloh, F.E. Inorganic Materials as Catalysts for Photochemical Splitting of Water. *Chem. Mater.* **2008**, *20*, 35–54. [CrossRef]

23. Nuida, T.; Kanai, N.; Hashimoto, K.; Watanabe, T.; Ohsaki, H. Enhancement of photocatalytic activity using UV light trapping effect. *Vacuum* **2004**, *74*, 729–733. [CrossRef]

24. Kajihara, K.; Yao, T. Macroporous Morphology of the Titania Films Prepared by a Sol-Gel Dip-Coating Method from the System Containing Poly(Ethylene Glycol). II. Effect of Solution Composition. *J. Sol-Gel Sci. Technol.* **1998**, *12*, 193–201. [CrossRef]

25. Grosso, D.; de A. A. Soler-Illia, G.J.; Babonneau, F.; Sanchez, C.; Albouy, P.A.; Brunet-Bruneau, A.; Balkenende, A.R. Highly Organized Mesoporous Titania Thin Films Showing Mono-Oriented 2D Hexagonal Channels. *Adv. Mater.* **2001**, *13*, 1085–1090. [CrossRef]

26. Roy, P.; Berger, S.; Schmuki, P. TiO_2 Nanotubes: Synthesis and Applications. *Angew. Chem. Int. Ed.* **2011**, *50*, 2904–2939. [CrossRef] [PubMed]

27. Park, J.T.; Roh, D.K.; Patel, R.; Kim, E.; Ryu, D.Y.; Kim, J.H. Preparation of TiO_2 spheres with hierarchical pores via grafting polymerization and sol–gel process for dye-sensitized solar cells. *J. Mater. Chem.* **2010**, *20*, 8521–8530. [CrossRef]

28. Chen, D.; Caruso, R.A. Recent Progress in the Synthesis of Spherical Titania Nanostructures and Their Applications. *Adv. Funct. Mater.* **2013**, *23*, 1356–1374. [CrossRef]

29. Ghosh, M.; Lohrasbi, M.; Chuang, S.S.C.; Jana, S.C. Mesoporous Titanium Dioxide Nanofibers with a Significantly Enhanced Photocatalytic Activity. *ChemCatChem* **2016**, *8*, 2525–2535. [CrossRef]

30. Ghosh, M.; Jana, S.C. Bi-component inorganic oxide nanofibers from gas jet fiber spinning process. *RSC Adv.* **2015**, *5*, 105313–105318. [CrossRef]

31. Liu, B.; Aydil, E.S. Growth of Oriented Single-Crystalline Rutile TiO_2 Nanorods on Transparent Conducting Substrates for Dye-Sensitized Solar Cells. *J. Am. Chem. Soc.* **2009**, *131*, 3985–3990. [CrossRef] [PubMed]

32. Ghosh, M.; Liu, J.; Chuang, S.S.C.; Jana, S.C. Fabrication of Hierarchical V_2O_5 Nanorods on TiO_2 Nanofibers and Their Enhanced Photocatalytic Activity under Visible Light. *ChemCatChem* **2018**, *10*, 3305–3318. [CrossRef]

33. Lee, K.; Mazare, A.; Schmuki, P. One-Dimensional Titanium Dioxide Nanomaterials: Nanotubes. *Chem. Rev.* **2014**, *114*, 9385–9454. [CrossRef] [PubMed]

34. Chandra, D.; Bhaumik, A. Super-microporous TiO_2 synthesized by using new designed chelating structure directing agents. *Micropor. Mesopor. Mat.* **2008**, *112*, 533–541. [CrossRef]

35. Lv, J.; Zhu, L. Highly efficient indoor air purification using adsorption-enhanced-photocatalysis-based microporous TiO_2 at short residence time. *Environm. Technol.* **2013**, *34*, 1447–1454. [CrossRef] [PubMed]

36. Fuertes, M.C.; Soler-Illia, G.J.A.A. Processing of Macroporous Titania Thin Films: From Multiscale Functional Porosity to Nanocrystalline Macroporous TiO_2. *Chem. Mater.* **2006**, *18*, 2109–2117. [CrossRef]

37. Yao, J.; Takahashi, M.; Yoko, T. Controlled preparation of macroporous TiO_2 films by photo polymerization-induced phase separation method and their photocatalytic performance. *Thin Solid Films* **2009**, *517*, 6479–6485. [CrossRef]

38. Malfatti, L.; Bellino, M.G.; Innocenzi, P.; Soler-Illia, G.J.A.A. One-Pot Route to Produce Hierarchically Porous Titania Thin Films by Controlled Self-Assembly, Swelling, and Phase Separation. *Chem. Mater.* **2009**, *21*, 2763–2769. [CrossRef]

39. Sun, W.; Zhou, S.; You, B.; Wu, L. Facile Fabrication and High Photoelectric Properties of Hierarchically Ordered Porous TiO_2. *Chem. Mater.* **2012**, *24*, 3800–3810. [CrossRef]

40. Chen, X.; Burda, C. The Electronic Origin of the Visible-Light Absorption Properties of C-, N- and S-Doped TiO_2 Nanomaterials. *J. Am. Chem. Soc.* **2008**, *130*, 5018–5019. [CrossRef]

41. Devi, L.G.; Kavitha, R. A review on plasmonic metal–TiO_2 composite for generation, trapping, storing and dynamic vectorial transfer of photogenerated electrons across the Schottky junction in a photocatalytic system. *Appl. Surf. Sci.* **2016**, *360*, 601–622. [CrossRef]

42. Islam, S.Z.; Nagpure, S.; Kim, D.Y.; Rankin, S.E. Synthesis and Catalytic Applications of Non-Metal Doped Mesoporous Titania. *Inorganics* **2017**, *5*, 15. [CrossRef]

43. Zaleska, A. Doped-TiO$_2$: A Review. *Recent Pat. Eng.* **2008**, *2*, 157–164. [CrossRef]

44. Hwang, Y.J.; Boukai, A.; Yang, P. High Density n-Si/n-TiO$_2$ Core/Shell Nanowire Arrays with Enhanced Photoactivity. *Nano Lett.* **2009**, *9*, 410–415. [CrossRef] [PubMed]

45. Qi, K.; Cheng, B.; Yu, J.; Ho, W. A review on TiO$_2$-based Z-scheme photocatalysts. *Chin. J. Catal.* **2017**, *38*, 1936–1955. [CrossRef]

46. Low, J.; Jiang, C.; Cheng, B.; Wageh, S.; Al-Ghamdi, A.A.; Yu, J. A Review of Direct Z-Scheme Photocatalysts. *Small Meth.* **2017**, *1*, 1700080. [CrossRef]

47. Feng, D.; Luo, W.; Zhang, J.; Xu, M.; Zhang, R.; Wu, H.; Lv, Y.; Asiri, A.M.; Khan, S.B.; Rahman, M.M.; et al. Multi-layered mesoporous TiO$_2$ thin films with large pores and highly crystalline frameworks for efficient photoelectrochemical conversion. *J. Mater. Chem. A* **2013**, *1*, 1591–1599. [CrossRef]

48. Chen, X.; Liu, L.; Huang, F. Black titanium dioxide (TiO$_2$) nanomaterials. *Chem. Soc. Rev.* **2015**, *44*, 1861–1885. [CrossRef]

49. Tian, Y.; Tatsuma, T. Plasmon-induced photoelectrochemistry at metal nanoparticles supported on nanoporous TiO$_2$. *Chem. Commun.* **2004**, 1810–1811. [CrossRef]

50. Bannat, I.; Wessels, K.; Oekermann, T.; Rathouský, J.; Bahnemann, D.; Wark, M. Improving the Photocatalytic Performance of Mesoporous Titania Films by Modification with Gold Nanostructures. *Chem. Mater.* **2009**, *21*, 1645–1653. [CrossRef]

51. Gellé, A.; Moores, A. Water splitting catalyzed by titanium dioxide decorated with plasmonic nanoparticles. *Pure Appl. Chem* **2017**, *89*, 1817–1827. [CrossRef]

52. Ghanem, M.A.; Arunachalam, P.; Amer, M.S.; Al-Mayouf, A.M. Mesoporous titanium dioxide photoanodes decorated with gold nanoparticles for boosting the photoelectrochemical alkali water oxidation. *Mater. Chem. Phys.* **2018**, *213*, 56–66. [CrossRef]

53. Couzon, N.; Maillard, M.; Chassagneux, F.; Brioude, A.; Bois, L. Photoelectrochemical Behavior of Silver Nanoparticles inside Mesoporous Titania: Plasmon-Induced Charge Separation Effect. *Langmuir* **2019**, *35*, 2517–2526. [CrossRef] [PubMed]

54. Primo, A.; Corma, A.; García, H. Titania supported gold nanoparticles as photocatalyst. *Phys. Chem. Chem. Phys.* **2011**, *13*, 886–910. [CrossRef] [PubMed]

55. Neațu, Ș.; Maciá-Agulló, J.A.; Concepción, P.; Garcia, H. Gold–Copper Nanoalloys Supported on TiO$_2$ as Photocatalysts for CO$_2$ Reduction by Water. *J. Am. Chem. Soc.* **2014**, *136*, 15969–15976. [CrossRef]

56. Jeong, S.; Kim, W.D.; Lee, S.; Lee, K.; Lee, S.; Lee, D.; Lee, D.C. Bi$_2$O$_3$ as a Promoter for Cu/TiO$_2$ Photocatalysts for the Selective Conversion of Carbon Dioxide into Methane. *ChemCatChem* **2016**, *8*, 1641–1645. [CrossRef]

57. Eustis, S.; El-Sayed, M.A. Why gold nanoparticles are more precious than pretty gold: Noble metal surface plasmon resonance and its enhancement of the radiative and nonradiative properties of nanocrystals of different shapes. *Chem. Soc. Rev.* **2006**, *35*, 209–217. [CrossRef]

58. Pan, J.H.; Zhao, X.; Lee, W.I. Block copolymer-templated synthesis of highly organized mesoporous TiO$_2$-based films and their photoelectrochemical applications. *Chem. Eng. J.* **2011**, *170*, 363–380. [CrossRef]

59. Mahoney, L.; Koodali, R.T. Versatility of Evaporation-Induced Self-Assembly (EISA) Method for Preparation of Mesoporous TiO$_2$ for Energy and Environmental Applications. *Materials* **2014**, *7*, 2697–2746. [CrossRef]

60. Sanchez, C.; Boissière, C.; Grosso, D.; Laberty, C.; Nicole, L. Design, Synthesis, and Properties of Inorganic and Hybrid Thin Films Having Periodically Organized Nanoporosity. *Chem. Mater.* **2008**, *20*, 682–737. [CrossRef]

61. Antonelli, D.M.; Ying, J.Y. Synthesis of a Stable Hexagonally Packed Mesoporous Niobium Oxide Molecular Sieve Through a Novel Ligand-Assisted Templating Mechanism. *Angew. Chem. Int. Ed.* **1996**, *35*, 426–430. [CrossRef]

62. Soler-Illia, G.J.A.A.; Angelome, P.C.; Fuertes, M.C.; Grosso, D.; Boissiere, C. Critical aspects in the production of periodically ordered mesoporous titania thin films. *Nanoscale* **2012**, *4*, 2549–2566. [CrossRef] [PubMed]

63. Li, W.; Wu, X.; Wang, J.; Elzatahry, A.A.; Zhao, D. A Perspective on Mesoporous TiO$_2$ Materials. *Chem. Mater.* **2014**, *26*, 287–298. [CrossRef]

64. Sánchez, V.M.; Martínez, E.D.; Martínez Ricci, M.L.; Troiani, H.; Soler-Illia, G.J.A.A. Optical Properties of Au Nanoparticles Included in Mesoporous TiO$_2$ Thin Films: A Dual Experimental and Modeling Study. *J. Phys. Chem. C* **2013**, *117*, 7246–7259. [CrossRef]

65. Martínez, E.D.; Boissière, C.; Grosso, D.; Sanchez, C.; Troiani, H.; Soler-Illia, G.J.A.A. Confinement-Induced Growth of Au Nanoparticles Entrapped in Mesoporous TiO$_2$ Thin Films Evidenced by in Situ Thermo-Ellipsometry. *J. Phys. Chem. C* **2014**, *118*, 13137–13151. [CrossRef]

66. Granja, L.P.; Martínez, E.D.; Troiani, H.; Sanchez, C.; Soler-Illia, G.J.A.A. Magnetic Gold Confined in Ordered Mesoporous Titania Thin Films: A Noble Approach for Magnetic Devices. *ACS Appl. Mater. Int.* **2017**, *9*, 965–971. [CrossRef]

67. Zalduendo, M.M.; Langer, J.; Giner-Casares, J.J.; Halac, E.B.; Soler-Illia, G.J.A.A.; Liz-Marzan, L.M.; Angelome, P.C. Au Nanoparticles-Mesoporous TiO$_2$ Thin Films Composites as SERS Sensors: A Systematic Performance Analysis. *J. Phys. Chem. C* **2018**, *122*, 13095–13105. [CrossRef]

68. Steinberg, P.Y.; Zalduendo, M.M.; Giménez, G.; Soler-Illia, G.J.A.A.; Angelomé, P.C. TiO$_2$ mesoporous thin film architecture as a tool to control Au nanoparticles growth and sensing capabilities. *Phys. Chem. Chem. Phys.* **2019**, *21*, 10347–10356. [CrossRef]

69. Li, H.; Shen, L.; Zhang, K.; Sun, B.; Ren, L.; Qiao, P.; Pan, K.; Wang, L.; Zhou, W. Surface plasmon resonance-enhanced solar-driven photocatalytic performance from Ag nanoparticle-decorated self-floating porous black TiO$_2$ foams. *Appl. Catal. B Environ.* **2018**, *220*, 111–117. [CrossRef]

70. Violi, I.L.; Perez, M.D.; Fuertes, M.C.; Soler-Illia, G.J.A.A. Highly Ordered, Accessible and Nanocrystalline Mesoporous TiO$_2$ Thin Films on Transparent Conductive Substrates. *ACS Appl. Mat. Int.* **2012**, *4*, 4320–4330. [CrossRef]

71. Brinker, C.J.; Hurd, A.J. Fundamentals of sol-gel dip-coating. *J. Phys. III France* **1994**, *4*, 1231–1242. [CrossRef]

72. Fuertes, M.C.; Colodrero, S.; Lozano, G.; González-Elipe, A.R.; Grosso, D.; Boissière, C.; Sánchez, C.; Soler-Illia, G.J.d.A.A.; Míguez, H. Sorption Properties of Mesoporous Multilayer Thin Films. *J. Phys. Chem. C* **2008**, *112*, 3157–3163. [CrossRef]

73. Krins, N.; Faustini, M.; Louis, B.; Grosso, D. Thick and Crack-Free Nanocrystalline Mesoporous TiO$_2$ Films Obtained by Capillary Coating from Aqueous Solutions. *Chem. Mater.* **2010**, *22*, 6218–6220. [CrossRef]

74. Procházka, J.; Kavan, L.; Shklover, V.; Zukalová, M.; Frank, O.; Kalbáč, M.; Zukal, A.; Pelouchová, H.; Janda, P.; Mocek, K.; et al. Multilayer Films from Templated TiO$_2$ and Structural Changes during their Thermal Treatment. *Chem. Mater.* **2008**, *20*, 2985–2993. [CrossRef]

75. Ismail, A.A.; Bahnemann, D.W.; Rathouský, J.; Yarovyi, V.; Wark, M. Multilayered ordered mesoporous platinum/titania composite films: Does the photocatalytic activity benefit from the film thickness? *J. Mater. Chem.* **2011**, *21*, 7802–7810. [CrossRef]

76. Koganti, V.R.; Dunphy, D.; Gowrishankar, V.; McGehee, M.D.; Li, X.; Wang, J.; Rankin, S.E. Generalized Coating Route to Silica and Titania Films with Orthogonally Tilted Cylindrical Nanopore Arrays. *Nano Lett.* **2006**, *6*, 2567–2570. [CrossRef] [PubMed]

77. Nagpure, S.; Das, S.; Garlapalli, R.K.; Strzalka, J.; Rankin, S.E. In Situ GISAXS Investigation of Low-Temperature Aging in Oriented Surfactant-Mesostructured Titania Thin Films. *J. Phys. Chem. C* **2015**, *119*, 22970–22984. [CrossRef]

78. Choi, S.; Mamak, M.; Coombs, N.; Chopra, N.; Ozin, G. Thermally Stable Two-Dimensional Hexagonal Mesoporous Nanocrystalline Anatase, Meso-nc-TiO$_2$: Bulk and Crack-Free Thin Film Morphologies. *Adv. Funct. Mater.* **2004**, *14*, 335–344. [CrossRef]

79. Vanýsek, P. Chapter 8: Preparation of Special Analytical Reagents—Electrochemical Series. In *CRC Handbook of Chemistry and Physics*; Lide, D.R., Ed.; CRC Press: Boca Raton, FL, USA, 2009; pp. 8.1–8.131.

80. Jiang, X.; Suzuki, N.; Bastakoti, B.P.; Chen, W.J.; Huang, Y.T.; Yamauchi, Y. Controlled Synthesis of Well-Ordered Mesoporous Titania Films with Large Mesopores Templated by Spherical PS-b-PEO Micelles. *Eur. J. Inorg. Chem.* **2013**, *2013*, 3286–3291. [CrossRef]

81. Bass, J.D.; Grosso, D.; Boissiere, C.; Sanchez, C. Pyrolysis, Crystallization, and Sintering of Mesostructured Titania Thin Films Assessed by in Situ Thermal Ellipsometry. *J. Am. Chem. Soc.* **2008**, *130*, 7882–7897. [CrossRef]

82. Djerdj, I.; Tonejc, A. Structural investigations of nanocrystalline TiO$_2$ samples. *J. Alloys Compd.* **2006**, *413*, 159–174. [CrossRef]

83. Swanson, H.E.; McMurdie, H.F.; Morris, M.C.; Evans, E.H. *Standard X-ray Diffraction Powder Patterns — Data for 81 Substances*; National Bureau of Standards: Washington, DC, USA, 1969; pp. 1–188.

84. Toledo-Antonio, J.; Gutiérrez-Baez, R.; Sebastian, P.; Vázquez, A. Thermal stability and structural deformation of rutile SnO_2 nanoparticles. *J. Solid State Chem.* **2003**, *174*, 241–248. [CrossRef]

85. Grosso, D.; Soler-Illia, G.J.d.A.A.; Crepaldi, E.L.; Cagnol, F.; Sinturel, C.; Bourgeois, A.; Brunet-Bruneau, A.; Amenitsch, H.; Albouy, P.A.; Sanchez, C. Highly Porous TiO_2 Anatase Optical Thin Films with Cubic Mesostructure Stabilized at 700 °C. *Chem. Mater.* **2003**, *15*, 4562–4570. [CrossRef]

86. Zhou, W.; Sun, F.; Pan, K.; Tian, G.; Jiang, B.; Ren, Z.; Tian, C.; Fu, H. Well-Ordered Large-Pore Mesoporous Anatase TiO_2 with Remarkably High Thermal Stability and Improved Crystallinity: Preparation, Characterization, and Photocatalytic Performance. *Adv. Funct. Mater.* **2011**, *21*, 1922–1930. [CrossRef]

87. Tao, J.; Sun, Y.; Ge, M.; Chen, X.; Dai, N. Non-Prefabricated Nanocrystal Mesoporous TiO_2-Based Photoanodes Tuned by A Layer-by-Layer and Rapid Thermal Process. *ACS Appl. Mater. Int.* **2010**, *2*, 265–269. [CrossRef]

88. Nagpure, S.; Zhang, Q.; Khan, M.A.; Islam, S.Z.; Xu, J.; Strzalka, J.; Cheng, Y.T.; Knutson, B.L.; Rankin, S.E. Layer-by-Layer Synthesis of Thick Mesoporous TiO_2 Films with Vertically Oriented Accessible Nanopores and Their Application for Lithium-Ion Battery Negative Electrodes. *Adv. Funct. Mater.* **2018**, *28*, 1801849. [CrossRef]

89. Wu, Q.L.; Subramanian, N.; Rankin, S.E. Hierarchically Porous Titania Thin Film Prepared by Controlled Phase Separation and Surfactant Templating. *Langmuir* **2011**, *27*, 9557–9566. [CrossRef]

90. Bicelli, L.P.; Bozzini, B.; Mele, C. A review of nanostructural aspects of metal electrodeposition. *Int. J. Electrochem. Sci.* **2008**, *3*, 356–408.

 *coatings*

Article

Fully Reversible Electrically Induced Photochromic-Like Behaviour of Ag:TiO$_2$ Thin Films

Stavros Katsiaounis [1], Julianna Panidi [1,2], Ioannis Koutselas [1,*] and Emmanuel Topoglidis [1,*]

1   Materials Science Department, University of Patras, 26504 Patras, Greece;
    stavroskatsiaounis@hotmail.com (S.K.); panidij@gmail.com (J.P.)
2   Department of Physics and Centre of Plastic Electronics, Imperial College London,
    South Kensington SW7 2AZ, UK
*   Correspondence: ikouts@upatras.gr (I.K.); etop@upatras.gr (E.T.); Tel.: +30-261-099-7727 (I.K.)

Received: 23 December 2019; Accepted: 28 January 2020; Published: 3 February 2020

Abstract: A TiO$_2$ thin film, prepared on fluorine-doped indium tin oxide (FTO)-coated glass substrate, from commercial off-the-shelf terpinol-based paste, was used to directly adsorb Ag plasmonic nanoparticles capped with polyvinylpyrrolidone (PVP) coating. The TiO$_2$ film was sintered before the surface entrapment of Ag nanoparticles. The composite was evaluated in terms of spectroelectrochemical measurements, cyclic voltammetry as well as structural methods such as scanning electron microscopy (SEM), transmission electron microscopy (TEM) and X-ray diffraction (XRD). It was found that the Ag nanoparticles are effectively adsorbed on the TiO$_2$ film, while application of controlled voltages leads to a fully reversible shift of the plasmon peak from 413 nm at oxidation inducing voltages to 440 nm at reducing voltages. This phenomenon allows for the fabrication of a simple photonic switch at either or both wavelengths. The phenomenon of the plasmon shift is due to a combination of plasmon shift related to the form and dielectric environment of the nanoparticles.

Keywords: TiO$_2$ films; Ag nanoparticles; optical properties; spectroelectrochemistry; cyclic voltammetry; surface plasmon

---

1. Introduction

In recent years there has been significant interest in optically transparent electrodes, due to their range of applications, including solar cells, light-emitting diodes and printable electronics [1,2]. Mesoporous (mp) nanocrystalline titanium dioxide (TiO$_2$) films are optically transparent for wavelengths greater than 390 nm due to their wide energy band gap, at ca. 3.2 eV [3]. The TiO$_2$ films comprise a rigid, porous network which is built with 10–40 nm nanocrystalline TiO$_2$ nanoparticles. These films usually exhibit pore sizes between 5 and 20 nm, sufficiently large for dye molecules [4], metal nanoparticles [5–8], biomolecules [9], gases [10], quantum dots [11,12] and perovskites [13] to diffuse throughout their porous structure. Their surface area is typically much greater (by up to 1000 times) than their geometric area [9]. In addition to their optical transparency and high surface area, these films exhibit good chemical stability, excellent optoelectronic properties and electrochemical activity at potentials above their conduction band edge. Therefore, TiO$_2$ films have been utilized in many applications such as photovoltaics [4], electrochromic windows and displays [14], antireflective coatings [13,15], batteries [14,16], touch screens [17], light-emitting diodes [18], supercapacitors [15,19], photocatalysis and photoelectrochemistry [20–23] or spectroelectrochemistry [9] applications, sensors [24–26] and biosensors [9].

The TiO$_2$ films are highly photosensitive and exhibit optically induced properties. They are also applicable in non-linear optics and optical devices [27,28]. However, one of their main disadvantages is that their large energy band gap (E_g = 3–3.4 eV), that lies in the ultraviolet (UV) region, limits their optical response in the visible region of the electromagnetic spectrum. In order to increase their absorption in the visible spectral range, the introduction of active absorbing units is required for optoelectronic devices. In fact, a number of studies over the last few years show that this can be achieved by incorporating noble metal nanoparticles with plasmonic effect in the TiO_2 matrix [27,29,30]. Plasmons are collective oscillations of electrons residing in unfilled energy bands and generally appear as pronounced resonances in the optical absorption spectra of metallic nanoparticles. This photochromic effect adds an interesting new aspect to the rich optical behaviour exhibited by silver nanoparticles (AgNPs). This effect can be studied among others in light scattering and absorption, non-linear signal enhancement and electroluminescence. Plasmonic nanostructures with increased and tunable optical absorption are used in various electronic devices, such as in thin solar cells through efficient scattering of the incident light in semiconducting absorber. Applications of plasmonic materials can also be considered for photochromic materials, which can reversibly change their colour under illumination or applied bias [31]. Lastly, plasmonic nanomaterials have also been proposed for a wide range of applications such as information storage and large-scale displays [32].

Multicolour photochromism was reported in nanocomposite Ag-TiO_2 films when these were prepared photocatalytically using a sol-gel route and consisted of AgNPs embedded in anatase TiO_2 [33]. The photochromic effect of the composites relied on burning a reversible spectral hole in the plasmon band.

Noble metal nanoparticles, such as AgNPs, exhibit an absorption band in the visible region of the spectrum caused by the surface plasmon resonance (SPR), which occurs at a different frequency from that of the bulk plasmon. The resonance wavelength of the SPR in AgNPs and its intensity are extremely dependent on the particle's environment (dielectric constant and interparticle distance) as well as on their geometry, size and shape. The incorporation of these nanoparticles into the TiO_2 matrix will extend their utility and device applications. Ag-TiO_2 films exhibit photochromic properties and, therefore, could be used for information storage, displays, smart windows and switches. Additional to these applications, the optically transparent semiconductor TiO_2 is used to carry out direct spectroelectrochemical experiments of molecules, such as redox proteins which were seen to be electrochemically active and changes in optical spectra were correlated with changes in applied potential [9].

Herein, we present the use of AgNPs as a simple electrically induced photonic switch when they are deposited on mesoporous (mp) TiO_2 films. The electrochromic behaviour of the Ag-TiO_2 nanocomposite film was characterized by cyclic voltammetry (CV) and spectroelectrochemistry (SEC). The crystallinity of the film was characterized by X-ray diffraction (XRD) and the surface morphology was examined by scanning electron microscopy (SEM) and transmission electron microscopy (TEM). The optical properties of the composite films were investigated via ultraviolet–visible (UV–Vis) spectroscopy. The optical properties and their morphology revealed a hybrid material whose plasmon can be tuned via the application of external bias. Furthermore, the fabrication of a simple photonic switch (on a rigid support) was attempted by assessing the electrochromic behaviour of the Ag-TiO_2 films by the application of controlled voltages using SEC. The novelty of the present work is based on the simple and straightforward preparation conditions of both the mp TiO_2 layer and AgNPs layer, which provide an optically interesting material. The SPR effect of AgNPs on mesoporous TiO_2 films could be influenced by charge transfer and local electric field enhancement. In the charge transfer mechanism, the SPR excites the electrons in the Ag nanoparticles, which are transferred to the conduction band of TiO_2, leaving a "plasmonic hole" in the metal nanoparticle [34].

2. Materials and Methods

2.1. Materials

Commercial 18NRT TiO_2 paste with average final nanoparticle size of 20 nm was purchased from Dyesol (Elanora, Australia) and used without any further purification. Fluorine doped tin oxide-coated (FTO) glass with resistance of 15 Ω/sq was purchased from Hartford Glass (IN, US). Sodium dihydrogen orthophosphate (0.01 M) was used to prepare the supporting electrolyte, and its pH was adjusted to 7 using NaOH. All other reagents were of chemical grade. All aqueous solutions were prepared in distilled, deionised water of resistance R = 18 MΩ cm. Silver nitrate (AgNO$_3$, MW:169.87) and polyvinylpyrrolidone (PVP, MW:10000) were supplied by Sigma Aldrich Chemie GmbH (Taufkirchen, Germany) as well as Na$_2$S in form of solid platelets. P25 nanotitania powder was commercially acquired from Degussa.

2.2. Mesoporous TiO$_2$ Film Electrodes Preparation

Dyesol TiO$_2$ nano-product (18NRT) was used to prepare thin TiO$_2$ films on FTO glass slides. The slides were first cleaned in a detergent solution using an ultrasonic bath for 15 min, followed by rinsing with de-ionised (DI) water and ethanol. TiO$_2$ was deposited on the substrate via the doctor-blade technique [9], by masking the glass substrates with tape which enabled the control of the thickness and the width of the area spread. The films were then allowed to dry for 20 min (evaporation of the solvent) before being sintered for 20 min at 450 °C. The resulting TiO$_2$ films were then cut in 1 cm^2 pieces.

2.3. P25 Film Preparation

P25-TiO$_2$ suspension was prepared from the P25 nanotitania powder, consisting of 80% anatase and 20% rutile which is manufactured by flame synthesis. An aqueous suspension of P25-TiO$_2$ was prepared by mixing 6 g of P25 TiO$_2$ powder, 60 µL of acetylacetone, 4 mL of water, 15 mL of ethanol, 1 mL of acetic acid and 60 µL of Triton X-100. The preparation of films on FTO glass is the same as with the Dyesol paste. For completeness, P25 have exhibited same properties to those prepared with the Dyesol product, with the exception that the Degussa powder leads to film with large scattering, thus, rendering it harder to measure its optical properties.

2.4. Preparation of Silver Nanoparticles (AgNPs) and TiO$_2$-AgNPs Films

We dissolved 0.01 g/mL of PVP at room temperature in triply distilled H$_2$O (10 mL), to which 100 µL of 0.5 M AgNO$_3$ solution were slowly added under stirring for one hour. The solution was kept for 24 hours after which its colour darkened. Similar reactions were achieved by an extra addition of equal volumes of 0.5 M Na$_2$S aqueous solution to the AgNO$_3$:PVP mixture. TiO$_2$ films were immersed in AgNO$_3$:PVP solution and kept for 24 hours. This procedure enables the reduction of Ag$^+$ [35], its agglomeration to nanoparticles as well as the prevention of further agglomeration to very large nanoparticles [36].

2.5. Film Characterization

The adsorption onto the TiO$_2$ films was monitored by recording the UV–Vis absorption spectra of the immobilized films at room temperature using a Shimadzu UV-1800 spectrophotometer. Contributions to the spectra from scatter and absorption by the TiO$_2$ film alone were subtracted by the use of AgNPs-free reference films. Prior to all spectroscopic measurements, the films were removed from the immobilization solution and rinsed in a buffer (methanol) solution several times to remove non-immobilized nanoparticles or excessive AgNO$_3$ solution. The photocatalytic process can be associated with the Ag-TiO$_2$ films, where ultra bandgap irradiation of titania generates an electron-hole pair, with possibility to reduce silver ions at the surface of the titania to silver metal. The conditions under experiments were free from UV irradiation. Typically, the Ag-TiO$_2$ prepared films appeared dark brown black in transmitted light. Chemical and thermal reduction have also been used to prepare Ag-TiO$_2$ films which, however, lacked the switching mechanism reported here.

XRD analyses of the TiO$_2$ films on FTO glass with or without AgNPs were performed using a Bruker D8 advance X-Ray Diffractometer (Bruker AXS GmbH, Karlsruhe, Germany) using a Cu Kα-radiation source set with an anode current of 40 mA and accelerating voltage of 40 kV with a scanning speed 0.015 degrees/second. The diffraction patterns were indexed by comparison with the Joint Committee on Powder Diffraction Standards (JCPDS) files number 21-1276 and 21-1272 for rutile and anatase respectively. The morphology and thickness of the TiO_2 film was analysed by a ZEISS EVO MA 10 SEM equipped with an energy-dispersive spectrometer (EDS, Oxford Instruments, 129 eV resolution). The thin films were in some cases sputtered with gold, of 5 nm thickness, in order increase the conductivity of the samples prior the SEM imaging. TEM studies of the TiO_2 nanoparticles, crystallites and Ag nanocrystals were carried out with a Philips CM20 electron microscope equipped with a Gatan GIF200 energy filter.

2.6. Electrochemical Measurements

Electrochemical and spectroelectrochemical experiments were performed using an Autolab PGStat 101 potentiostat. The spectroelectrochemical cell was a 6 mL, three-electrode teflon cell with quartz windows, employing a platinum mesh flag as the counter electrode, an $Ag/AgCl/KCl_{sat}$ reference electrode, and the Ag-TiO_2 film on FTO conducting glass as the working electrode. The electrochemical cell had an inlet and an outlet for passing gas into it. The electrolyte, an aqueous solution of 10 mM sodium phosphate (pH 7), was thoroughly de-aerated by bubbling with Argon prior to any electrochemical measurements and an Argon atmosphere was maintained throughout the measurements. For spectroelectrochemistry, the above cell was incorporated in the sample compartment of the Shimadzu UV-1800 spectrophotometer, and the absorption changes were monitored as a function of the applied potential. All potentials are reported against Ag/AgCl and all experiments were carried out at room temperature.

3. Results and Discussion

The surface morphology, structure and thickness of the TiO_2 films prior and after deposition of AgNPs were analysed by SEM. The SEM images (Figure 1a) of TiO_2 film showed disordered porosity and comprise a rigid, porous network of TiO_2 nanoparticles of average size 30–40 nm. The film exhibits great homogeneity and even size distribution, while all nanoparticles are bonded together through the sintering process, creating a rich mesoporous surface. The thickness of the coated titania films (Figure 1b) was found to be around 6 μm by analysing the cross-section SEM images. These results confirm that the mesoporous film structure of TiO_2 could provide an excellent surface for the AgNPs to diffuse throughout the porous structure. In addition, the porous film could provide many active sites for electrocatalytic reactions. Mesoporous layers are most suitable for immobilizing electroactive compounds, as the surface area available for sensitization and hence electrochemistry can be increased by over two orders of magnitude with respect to a flat electrode, while ensuring satisfactory access to the pores [37]. Agglomerates (showed with a red arrow in Figure 1c) are formed when AgNPs are deposited on top of the TiO_2 films. The AgNPs agglomerates are shown as white dots on the SEM images at the sample surface and exhibit a large size distribution from 200 to 800 nm, with an average size of 600 nm. Back-scattered electrons (BSE) images have also provided evidence towards the different chemical content of the agglomerates with respect to that of the surface. The EDS spectrum of Ag-TiO_2 (Figure 1d) shows the main peaks of Ti, O, Ag confirming the large amount of silver present. The extra addition of equal volumes of 0.5 M Na_2S in the $AgNO_3$:PVP mixture, which most probably would have passivated the surface of the AgNPs with a thin layer of Ag_2S, was not detected in the EDS spectra, yet it is considered that a thin layer of Ag_2S may have formed on the AgNPs surface. Finally, EDS elemental analysis led to the conclusion that the TiO_2:Ag molar ratio at the surface was 8, while in the Supplementary Information (SI Figure S1) a surface EDS mapping can be observed that indicates the successful Ag coverage.

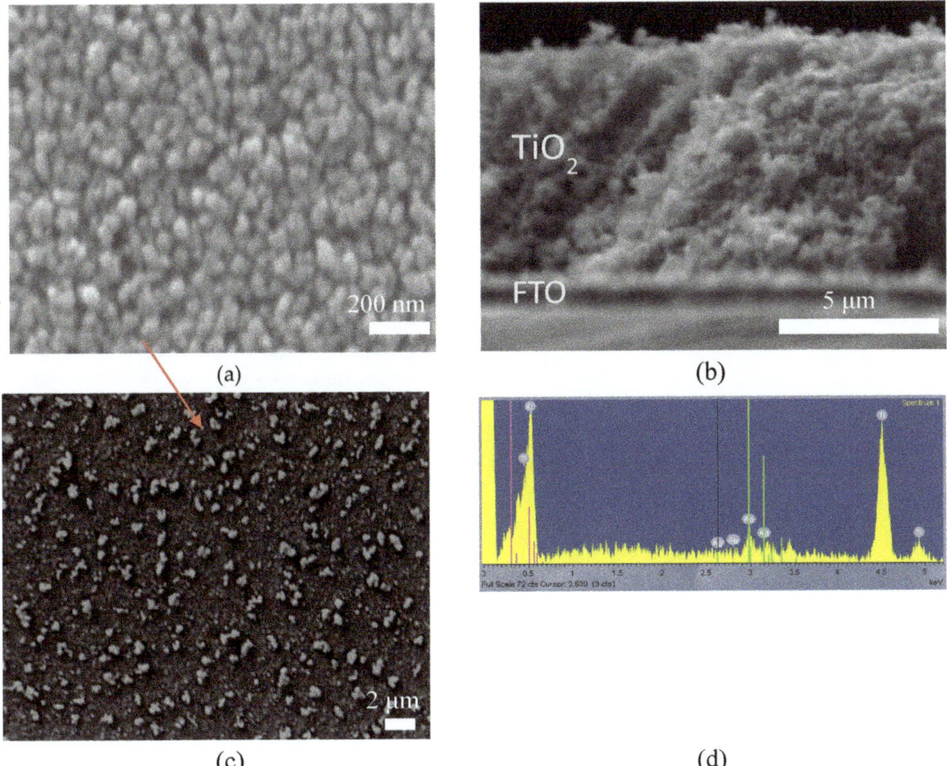

Figure 1. Scanning electron microscope (SEM) images for a bare TiO_2 film on fluorine-doped indium tin oxide (FTO) substrate (**a**) top view and (**b**) cross section, (**c**) SEM image of Ag-TiO_2 film and (**d**) energy-dispersive spectrometry (EDS) of Ag-TiO_2 on FTO substrate.

TEM was employed on Ag-TiO_2 films in order to further investigate the surface morphology and the size of the AgNPs. TiO_2 nanoparticles (Figure 2a,b) exhibit the shape of platelets with external large dimension of 20–45 nm, in agreement with the specifications of the Dyesol and/or P25 products. AgNPs (Figure 2c,d) form different structures; in some cases plate-like, when they are deposited on TiO_2 films with average diameter of 15 nm. High-resolution TEM (HR-TEM) imaging, Figure 2c, permitted easy differentiation of Ag nanocrystals (small dark areas) and TiO_2 crystallites (large bright areas). Ag nanocrystals can be observed on the surface of the TiO_2 particles as dark lines. It is presumed that even smaller Ag nanoparticles are in existence scattered throughout the porous TiO_2 surface.

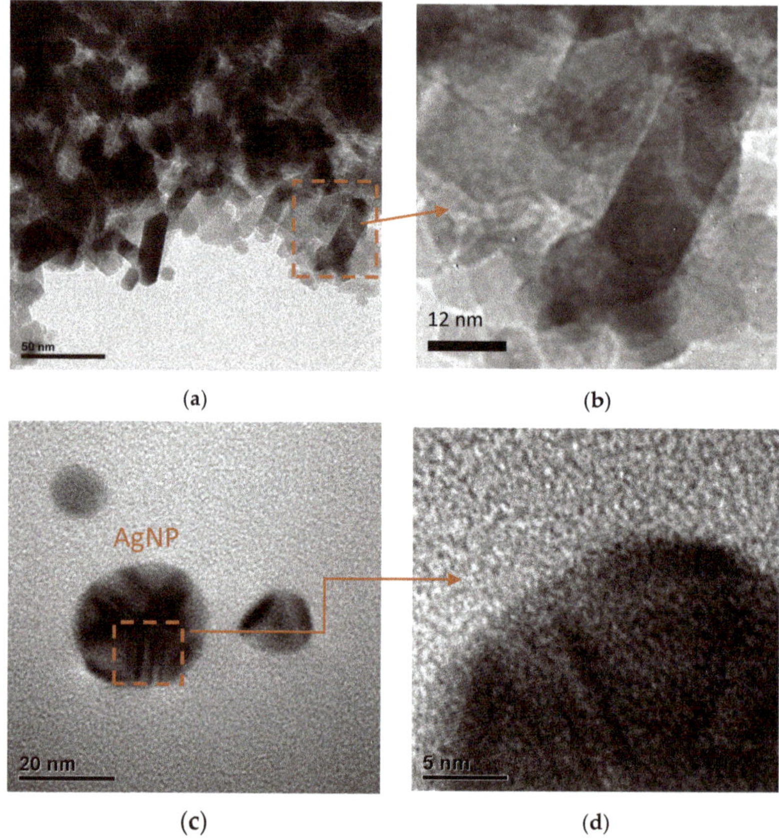

(a)

(b)

(c)

(d)

Figure 2. Transmission electron microscope (TEM) images (**a**,**b**) of TiO_2 at low TEM resolution and high-resolution TEM (HR-TEM) images (**c**,**d**) of Ag-TiO_2 composites. Images, a and c, on the left have been magnified at the selected dotted squares and placed to the right as b and d, respectively.

The color change of the Ag-TiO_2 films is significant upon bias application. Figure 3 presents the thin films of TiO_2 as prepared from Dyesol (Figure 3a) and from Degussa (Figure 3d). After immersing the TiO_2 films in the AgNPs solution the film color changes (Figure 3b), since AgNPs were adsorbed on the TiO_2 film. In order to perform the electrochemical measurements, negative bias was applied in the film and the color of the film became darker (Figure 3c,e).

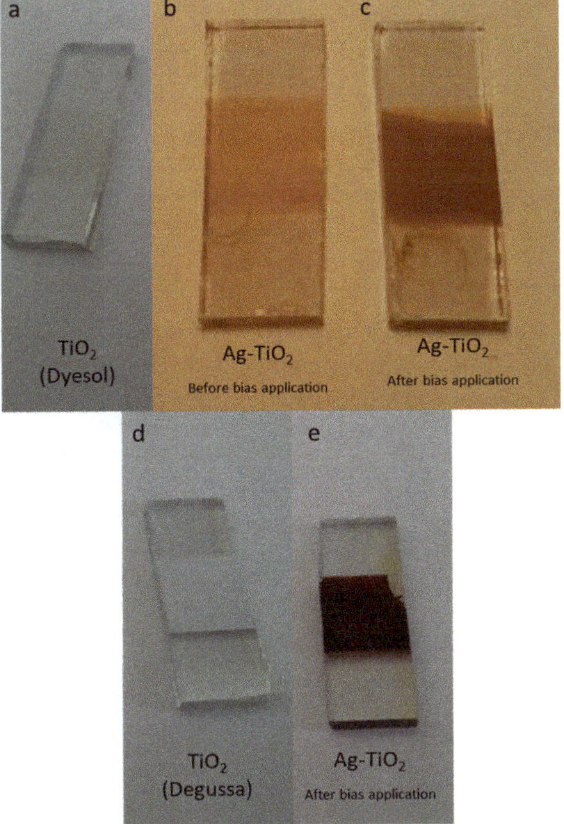

Figure 3. Digital photos (**a**) mesoporous TiO_2 film as fabricated from Dyesol and (**d**) from Degussa precursors. Ag-TiO_2 films (**b**) before and (**c,e**) after bias application.

In order to investigate any changes in the crystal structure of the TiO_2 films affected by the AgNPs, the XRD patterns (Figure 4) of FTO-conducting substrates, TiO_2 and Ag-TiO_2 film electrodes were measured. The XRD of TiO_2 and AgNPs-TiO_2 electrodes revealed similar characteristic peaks at 2θ: 25.28°, 37.8° and 48.05° which correspond to the indices of anatase TiO_2 (101), (004) and (200) hkl planes and are consistent with the reported values of the JCPDS file (21-1272). As expected, no characteristic peaks that correspond to the rutile TiO_2 indices were observed since the Dyesol TiO_2 paste is 100% anatase. The only difference was observed on the relative intensities of the peaks, which may be due to the fact that doping alters the crystallinity but not the crystal structure of the Ag-TiO_2 films. In addition, two slightly intense peaks at 26.54° and 38° which correspond to the plane indices 110 and 200 of the FTO glass could be observed. The XRD pattern of the AgNPs-TiO_2 film showed a new peak at 44.4° which can be assigned to the (200) plane of Ag. Furthermore, a closer look of the peak at 38° of the Ag-TiO_2, which corresponds to the (200) FTO index, also showed a small shoulder at 38.1° which can be assigned to the Ag (111) plane. These two peaks agree with the JCPDs card of Ag as presented in Figure 4. No rutile phase or any other modification is observed for the Ag-TiO_2 film depending on AgNPs incorporation.

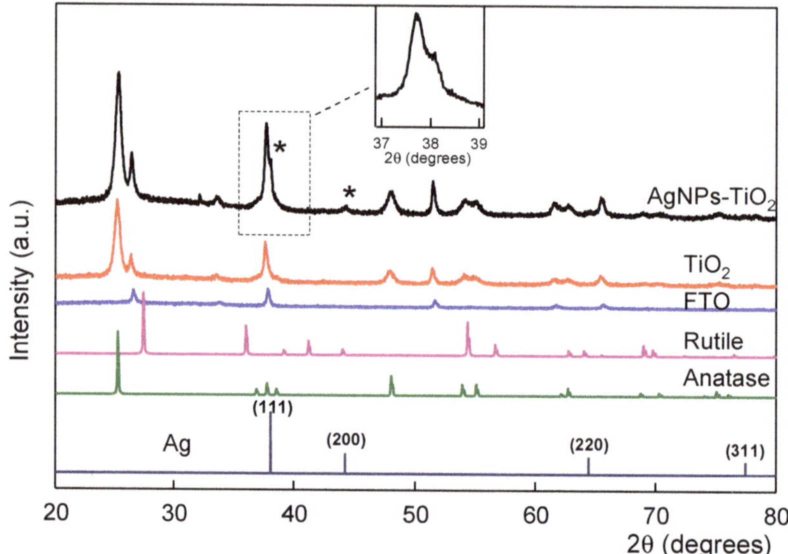

Figure 4. X-ray diffraction (XRD) patterns of a TiO$_2$ film with or without silver nanoparticles (AgNPs) deposited on its surface, FTO glass and Joint Committee on Powder Diffraction Standards (JCPDS) cards of anatase and rutile TiO$_2$. Inset: magnification of the peak at 38° corresponding to the fcc Ag (111) plane.

The TiO$_2$ films deposited on conducting FTO glass slides had been soaked in solutions of AgNPs and rinsed with NaH$_{2-}$PO$_4$ buffer to remove any loosely bound nanoparticles. The mp TiO$_2$ films combine transparency in the visible region of the electromagnetic spectrum with a high surface area accessible to molecules from a surrounding solution (e.g., AgNPs). In many cases in the past, this allowed the adsorbed molecules (dyes, proteins, electrochromic species and surfactants) to achieve the densities required for informative electronic absorption spectroscopy, whether that was for the development of solar cells, electrochemical biosensors, electrochromic devices or catalytic applications [3,4,9,27,38,39]. Therefore, the transparency of the electrodes can be useful as it could allow the AgNPs adsorption process and their electrochromic properties to be monitored by UV–Vis absorption spectroscopy, which is a technique also used by other authors for the structural characterization of the AgNPs in a dielectric matrix [27,38–41].

Figure 5 shows the optical absorption spectra of the AgNPs solution used as dopants for the surface of the TiO$_2$ films. A relatively broad absorption band is located at 452 nm properly corresponding to the size and form of the AgNPs; this peak is due to the SPR of the AgNPs and is within the spectrum range (400–450 nm) reported for them depending on their shape and size [27,30,38,39,41]. Adsorption of AgNPs on TiO$_2$ films results in light coloration (see Figure 3) of the films, indicating that a large amount of AgNPs has been immobilized into the mesoporous TiO$_2$ film. Also shown is the absorption spectrum of a bare TiO$_2$ film which is transparent and colourless in the visible region, showing a characteristic absorption increase below 400 nm due to the onset of TiO$_2$ band gap excitation. Therefore, the optical transparency of the TiO$_2$ allows the adsorption process of the AgNPs to be monitored by UV–Vis absorption [30,42]. The electronic absorption spectra (Figure 5) for the nanocomposite Ag-TiO$_2$ film showed features typical of AgNPs superimposed on a background arising from scattering by the TiO$_2$ layer. The resulting spectrum of AgNPs on the TiO$_2$ film showed a characteristic absorption band at 413 nm (a clear blue shift in comparison with the absorption spectra of the AgNPs in solution). The Ag-TiO$_2$ film exhibits a much narrower, blue shift and more defined optical spectrum which means that the active metallic cores of the deposited AgNPs are smaller than those in the solution and probably spherical [27,38,41,43]. The shift is related to the size of the AgNPs and to the interaction between TiO_2 and Ag, as well as to the fact that the majority of the AgNPs that have filled the pores are smaller than the pores themselves. No band is observed of organic residuals, which were used to bind the TiO2 nanoparticles, remaining in the film due to the titania film sintering at 450 °C. If the observed plasmonic bands were broad between 510 and 590 nm, that would have implied that the nanoparticles are large and/or of non spherical nature [44]. Increasing the $AgNO_3$ solution concentration, where the AgNPs were created, had as a result the increase of the absorbance of the Ag-TiO_2 film, as was also reported by other researchers in the past [30]. There is also the possibility that some of the non-reduced silver ions (Ag^+) could be adsorbed on the hydroxylated TiO_2 surface according to the following reaction, as suggested by other authors also in the past: $Ag^+ + TiO_2^- \rightarrow Ti$–$O$–$Ag + H^+$ [45].

The plasmon resonance in the Ag-TiO_2 thin films strongly depends on the crystalline phase and dielectric constant of the TiO_2 matrix [46]. In theory, the SPR peak wavelength increases with increasing dielectric constant of the matrix and depends on the refractive index.

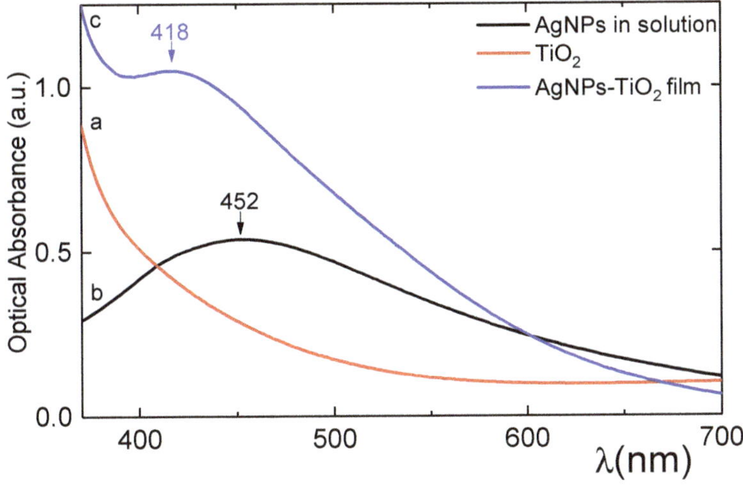

Figure 5. Optical absorption spectra in the visible range (380–700 nm) of (a) AgNPs solution, (b) TiO_2 film and (c) Ag-TiO_2 film.

It should be mentioned that the form of the absorption profile does not only signify that the 415 nm absorption peak of the Ag-TiO_2 films is characteristic of spherical AgNPs. It also carries information due to the band absence between 520 and 540 nm which is usually related to bigger AgNPs or AgNPs dimmers. Similarly, no band is observed around 620 nm which is usually due to non-spherical Ag nanoparticles or to larger particles or to manifestation of higher-order plasmon modes called quadrupolar modes. No band is also observed around 670 nm which suggests a longitudinal plasmon band of Ag nanorods. Any broadness of the observed absorption bands could refer to different morphologies of the deposited nanoparticles, to broad particle size distribution, and to agglomeration processes, while all the above discussion is in agreement with the TEM analysis.

The optical band gap of the Ag-TiO_2 films is expected to be smaller to that of pure (blank) TiO_2 films due to the effect of the AgNPs, as decrease of E_g with silver addition has also been reported by other authors [27,38,39,41]. Also, the Ag^+ ions probably exist on the surface of the anatase TiO_2 films by forming Ag–O–Ti bonds [47], which may introduce trap states affecting the energy band gap. Finally, it is possible that the potential applied can oxidize or reduce the Ag nanoparticles to silver oxide and reverse this as suggested by Kuzma et al [48].

Figure 6 presents the spectroelectrochemical spectrum for a Ag-TiO_2 film electrode after remaining for 2 min at each negative applied potential (−0.1 to −1.1 V) applied. The plasmon (or Soret) band at

413 nm up until the application of −0.4 V remains constant, but upon the application of −0.5 V, it starts to increase in intensity without though changing shape. The application of −0.6 V to the film causes a big increase of the peak which at the same time becomes thinner and sharper. Afterwards, and by gradually applying more negative biases up to −1.1 V, the peak at 413 nm continues to rise and at the same time becomes sharper. However, several minutes after the end of the spectroelectrochemical measurements and the application of no voltage, the absorbance of the film shows that it has not returned to its initial state, but rather a shift of the main peak from 413 nm to 440 nm has been observed.

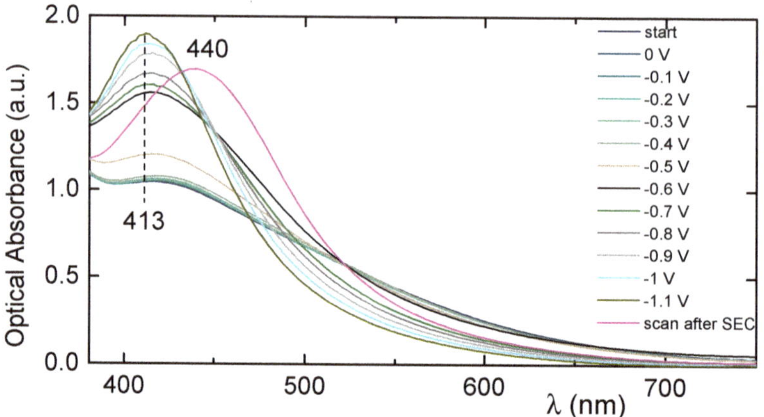

Figure 6. Ultraviolet–visible (UV–Vis) spectral changes of a Ag-TiO$_2$ film electrode upon the application of increasingly negative potentials (0 to −1.1 V vs. Ag/AgCl).

Following this route, increasing negative biases (−0.2 to −0.9) were applied again to the doped film and its absorption spectra were recorded. Figure 7 shows that upon the application of up to −0.4 V, a small increase in the size of the peak at 440 nm was monitored and a slight blue shift. However, upon the application of higher negative biases (−0.8 or −0.9 V) the peak shifts back to 413 nm. Figure 7 also shows that upon the application of a positive bias (0 to 0.6 V) the peak shifts again to 440 nm. The switch from 413 nm to 440 nm, depending on the bias that was applied to the film, was repeated several times and always with success.

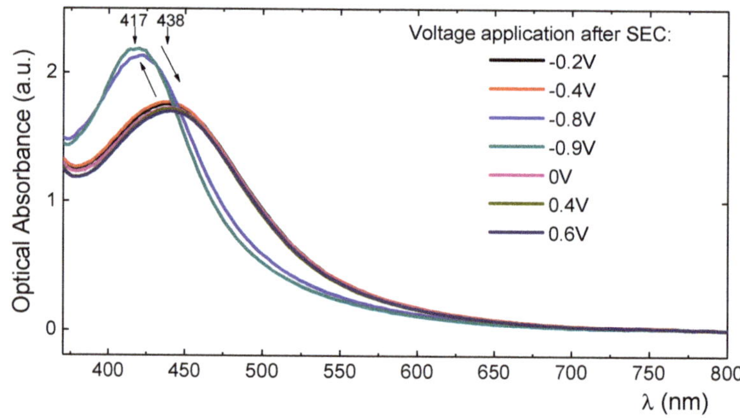

Figure 7. UV–Vis spectral changes of a Ag-TiO$_2$ film electrode under the application of negative or positive potentials showing the shift of the Soret peak from 413 nm to 440 nm and vice versa.

In order to gain an insight into the kinetic mechanism, the optical absorbance (OA) or optical density (OD) were measured for a range of negative and positive voltages. Figure 8a,b show the curves regarding the kinetics for the absorption change at 413 (a) and 440 nm (b) respectively. The absorbance was initially monitored during the application of −0.8 V for 600 s and immediately afterwards upon the application of 0.15 V for another 600 s. Figure 8a demonstrates that the continuous application of a sufficient cathodic current (−0.8 V) causes the fast increase of the absorbance at 413 nm of the Ag-TiO$_2$ film in 4.5 s ($\tau_{1/2}$ = 2.6 s). By stepping back the potential from −0.8 V to +0.15 V, the absorbance at 413 nm starts dropping quite fast at the first 20 s and slower afterwards until it reaches the OD value it exhibited before the application of the negative bias. Similar results were obtained in Figure 8b for the kinetics for the absorption change at 440 nm. The continuous application of the −0.8 V caused the fast increase of OD at 440 nm of the Ag-TiO$_2$ in 6 sec ($\tau_{1/2}$ = 3.8 s). The application of the +0.15 V caused the gradual drop of the OD at 440 nm and reached completion after 50 s.

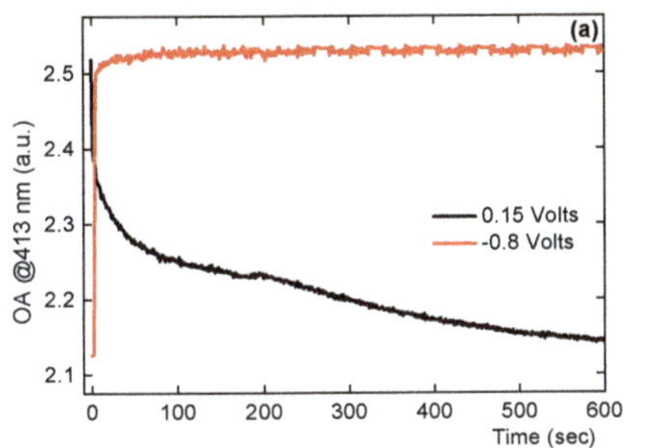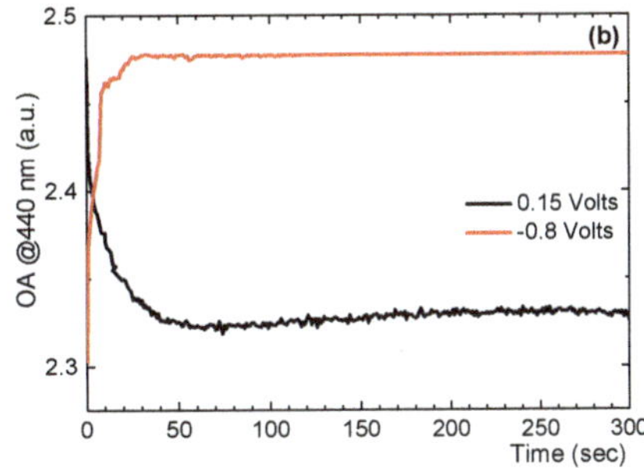

Figure 8. Kinetics for the absorption change at 413 nm (**a**) and 440 nm (**b**) of a 6 μm thick Ag-TiO$_2$ film by the application of 0.15 V and −0.8 V.

The spectroelectrochemical studies reported in Figures 6 and 7 were further supported by cyclic voltammetry (CV). The CVs of mp TiO$_2$ films with or without the adsorption of AgNPs on their surface, in aqueous 10 mM NaH$_2$PO$_4$ electrolyte of pH 7, at a scan rate of 0.1 V/s, are presented in Figure 9. Upon scanning the potential of a bare semiconductive TiO$_2$ film cathodically in a neutral pH aqueous medium (NaH$_2$PO$_4$, pH 7), a transition from insulating to conductive behaviour is observed. The characteristic charging/discharging currents were assigned to electron injection and storage into sub-band gap/conduction band states of the TiO$_2$ film, until the metal oxide becomes fully degenerated once the applied potential reaches the conduction band potential [9,33,34,49]. The current shows a plateau at potentials where the film behaves as an insulator (positive biases and up to around −0.3 V). At that range the electrical response is dominated by the Helmholtz capacity of the uncovered FTO glass/electrolyte solution interface at the bottom of the TiO$_2$ film [50,51]. At more negative potentials (−0.3 V and higher), the cathodic current displays an exponentially rising behaviour (at −0.8 V) that is considered by many authors as the reduction of superficial Ti ions [52,53]. This rising behaviour is then transformed to a peak (at −0.75 V), when the direction of voltammetry is reversed to anodic. This is considered as re-oxidation of reduced Ti ions and is a slow and irreversible process [53]. These potentials are more negative than the conduction band potential and the TiO$_2$ film behaves as a conductive metallic electrode. As the scan rate becomes slower (Figure 10) the height of the anodic peak progressively diminishes until, at very slow scan rates (0.01 V/s), it disappears. According to many reports [1,7,53,54] the origin of this behaviour is due to the charging/discharging of electrons in the film and a charge transfer mechanism. However, no cathodic or anodic peaks due to a redox reaction are observed. The CV integrates to approximately 0, indicating negligible Faradaic currents.

In addition, Figure 9 shows the CV of a TiO$_2$ film electrode after the adsorption of AgNPs on its surface. The Ag-TiO$_2$ film exhibited two pairs of redox peaks that were absent from the CV of the bare TiO$_2$ film. For the first pair, a small cathodic peak is observed at +0.07 V and a small anodic peak at

+0.27 V. For the second pair the cathodic peak is much bigger and appears at −0.65 V and the anodic peak which is much smaller and broader appears around −0.53 V. All these peaks are attributed to two different reduction/oxidation states of Ag on the TiO_2 films. The redox reactions are associated with $Ag \rightarrow Ag_2O$ and $Ag_2O \rightarrow Ag$ (oxidation and reduction peaks, respectively) [55,56]. The pair of peaks at the lower biases correspond to the reduction and oxidation of the AgNPs that have adsorbed inside the mesoporous network of the TiO_2 film and the other pair of smaller peaks most probably correspond either to small deposited AgNPs or to AgNPs adsorbed only on the outer surface of the TiO_2 film [46]. Also both re-oxidation peaks are broader and smaller to the reduction peaks due to the large band gap of the TiO_2 film. This observation most probably requires further experiments in order to draw any final conclusions. However, this redox behaviour (two sets of redox peaks, a two phase reduction) of the AgNPs on the TiO_2 films has also been monitored on immobilization studies of a heme based redox dyes, such as iron(III) 5,10,15,20-tetrakis(1-methyl-4-pyridyl) porphyrin, FeTMPyP, on the same films [33].

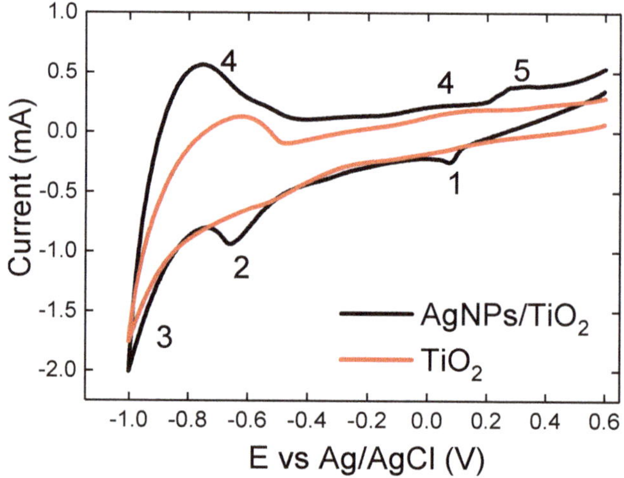

Figure 9. Cyclic voltammograms (CVs) in 10 mM NaH_2PO_4 buffer, pH 7 for Ag-TiO_2 and blank TiO_2 films. The scan rate was 0.1 V/s.

The effect of the scan rate on the voltammetric behaviour of the Ag-TiO_2 film was also investigated. Slower scan rates were applied to the Ag-TiO_2 electrode in order to try to obtain a reversible peak-shaped CV. Figure 10 illustrates that even at slow scan rates no simple reversible behaviour for the Ag-TiO_2 film was observed, consistent with the currents being limited by the low TiO_2 conductivity at moderate potentials. However, the cathodic peak potential at −0.65 V shifted negatively with the increase of the scan rate. In addition, there is a good linear relationship between the cathodic peak current and the scan rate in the range (0.01 to 0.1 V/s), indicative of a surface-confined electrochemical process. This has also been observed in other systems where instead of AgNPs, surfactants or proteins were adsorbed on the TiO_2 films [39].

Also, it was found that the cathodic and anodic peak currents corresponding to the adsorbed AgNPs vary linearly with scan rates in the range of 10 to 100 $mV \cdot s^{-1}$. The CVs exhibit a current increase as the scan rates are steadily increased, significantly larger when the 0.1 V/s scan rate results are compared to those of the 0.01 V/s. This is clear indication that the adsorbed AgNPs underwent a surface-controlled process.

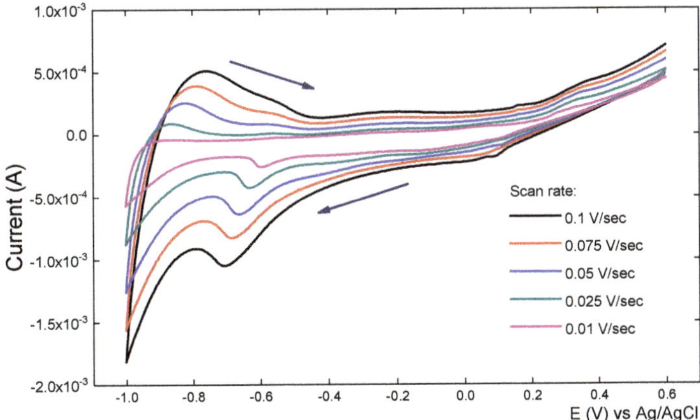

Figure 10. Cyclic voltammograms (CVs) in 10 mM NaH_2PO_4 buffer pH 7, for a Ag-TiO_2 film at different scan rates.

4. Conclusions

All the above experimental data clearly suggest that the electrophotochromism of the Ag-TiO_2 composite is due to oxidation/reduction of the Ag nanoparticles, which forms a thin layer of Ag_2O on the metallic core, forming core/shell nanoparticles; the core is metallic while the outer shell is semiconducting, where the peaks agree well with those found by Kuzma et al. [46]. In fact, it is suggested that the oxide thickness is about 1.5 nm when the composite absorbs at the 440 nm peak and at most 0.5 nm while the composite exhibits a peak at 413 nm. Furthermore, it is possible that the particles may be charged upon the application of negative potential on the Ag-TiO_2 electrode, which would lead their separation due to their charging; such an effect would not be possible at positive electrode potentials since silver has a lower conduction band than that of the FTO electrode and than that of TiO_2.

Supplementary Materials: The following are available online at http://www.mdpi.com/2079-6412/10/2/130/s1.

Author Contributions: E.T. and I.K. conceived and designed the experiments; S.K., J.P., E.T. and I.K. performed the experiments; E.T., I.K., S.K. and J.P. analyzed the data; E.T. and I.K. contributed reagents/materials/analysis tools; I.K., S.K., E.T. and J.P. wrote the paper. All authors have read and agreed to the published version of the manuscript.

Funding: This research received no external funding from funding agencies in the public, commercial or not-for-profit sectors.

Acknowledgments: We would like to thank Nikolaos Boukos from NCSR "Demokritos" for the TEM images.

Conflicts of Interest: The authors declare no conflict of interest.

References

1. Garoz-Ruiz, J.; Ibañez, D.; Romero, E.C.; Ruiz, V.; Heras, A.; Colina, A. Optically transparent electrodes for spectroelectrochemistry fabricated with graphene nanoplatelets and single-walled carbon nanotubes. *RSC Adv.* **2016**, *6*, 31431–31439. [CrossRef]

2. Ellmer, K. Past achievements and future challenges in the development of optically transparent electrodes. *Nat. Photonics* **2012**, *6*, 809–817. [CrossRef]

3. Topoglidis, E.; Lutz, T.; Durrant, J.R.; Palomares, E. Interfacial electron transfer on cytochrome-c sensitized conformally coated mesoporous TiO_2 films. *Bioelectrochemistry* **2008**, *74*, 142–148. [CrossRef] [PubMed]

4. O'Regan, B.; Gratzel, M. A low cost high-efficiency solar-cell on dye-sensitized colloidal TiO_2 films. *Nature* **1991**, *353*, 3737–3740. [CrossRef]

5. Xin, B.F.; Ren, Z.Y.; Hu, H.Y.; Zhang, X.Y.; Dong, C.L.; Shi, K.Y.; Jing, L.Q.; Fu, H.G. Photocatalytic activity and interfacial carrier transfer of Ag-TiO$_2$ nanoparticle films. *Appl. Surf. Sci.* **2005**, *252*, 2050–2055. [CrossRef]

6. Sarkar, J.; John, V.T.; He, J.; Brooks, C.; Gandhi, D.; Nunes, A.; Ramanath, G.; Bose, A. Surfactant-templated, synthesis and catalytic properties of patterned nanoporous titania supports loaded with platinum nanoparticles. *Chem. Mater.* **2008**, *20*, 5301–5306. [CrossRef]

7. Rodriguez, J.A.; Evans, J.; Graciani, J.; Park, J.B.; Liu, P.; Hrbek, J.; Sanz, J.F. High water-gas shift activity in TiO$_2$(110) supported Cu and Au nanoparticles: Role of the oxide and metal particle size. *J. Phys. Chem. C* **2009**, *113*, 7364–7370. [CrossRef]

8. Zhao, G.L.; Kozuka, H.; Yoko, T. Sol-gel preparation and photoelectrochemical properties of TiO$_2$ films containing Au and Ag metal particles. *Thin Solid Film.* **1996**, *277*, 147–154. [CrossRef]

9. Topoglidis, E.; Campbell, C.J.; Cass, A.E.G.; Durrant, J.R. Factors that affect protein adsorption on nanostructured titania films, A novel spectroelectrochemical application to sensing. *Langmuir* **2001**, *17*, 7899–7906. [CrossRef]

10. Ha, T.-J.; Hong, M.-H.; Park, C.-S.; Park, H.-H. Gas sensing properties of ordered mesoporous TiO$_2$ film enhanced by thermal shock induced cracking. *Sens. Actuat. B Chem.* **2013**, *181*, 874–879. [CrossRef]

11. Feng, J.; Han, J.; Zhao, X. Synthesis of CuInS$_2$ quantum dots on TiO$_2$ porous films by solvothermal method for absorption layer of solar cells. *Prog. Org. Coat.* **2009**, *64*, 268–273. [CrossRef]

12. Gao, X.F.; Li, H.B.; Sun, W.T.; Chen, Q.; Tang, F.Q.; Peng, L.M. CdTe quantum dots-Sensitized TiO$_2$ nanotube array photoelectrodes. *J. Phys. Chem. C* **2009**, *113*, 7531–7535. [CrossRef]

13. Giordano, F.; Abate, A.; Correa Baena, J.P.; Saliba, M.; Matsui, T.; Im, S.H.; Zakeeruddin, S.M.; Nazeeruddin, M.K.; Hagfeldt, A.; Graetzel, M. Enhanced electronic properties in mesoporous TiO$_2$ via lithium doping for high-efficiency perovskite solar cells. *Nat. Commun.* **2016**, *7*, 10379. [CrossRef] [PubMed]

14. Cinnsealach, R.; Boschloo, G.; Rao, S.N.; Fitzmaurice, D. Coloured electrochromic windows based on nanostructured TiO$_2$ films modified by adsorbed redox chromophores. *Sol. Energy Mater. Sol. Cells* **1999**, *57*, 107–125. [CrossRef]

15. Zhang, X.; Fujishima, A.; Jin, M.; Emeline, A.V.; Murakami, T. Double-layered TiO$_2$-SiO$_2$ nanostructured films with self-cleaning and antireflective properties. *J. Phys. Chem. B* **2006**, *110*, 25142–25148. [CrossRef]

16. Singhal, A.; Skandan, G.; Amatucci, G.; Badway, F.; Ye, N.; Manthiram, A.; Ye, H.; Xu, J.J. Nanostructured electrodes for next generation rechargeable electrochemical devices. *J. Power Sources* **2004**, *129*, 38–44. [CrossRef]

17. Hecht, D.S.; Hu, L.; Irwin, G. Emerging transparent electrodes based on thin films of carbon nanotubes, graphene, and metallic nanostructures. *Adv. Mater.* **2011**, *23*, 1482–1513. [CrossRef]

18. Cao, W.; Li, J.; Chen, H.; Xue, J. Transparent electrodes for organic optoelectronic devices: A review. *J. Photonics Energy* **2014**, *4*, 40990. [CrossRef]

19. Weng, Z.; Guo, H.; Liu, X.; Wu, S.; Yeung, K.; Chu, P.K. Nanostructured TiO$_2$ for energy conversion and storage. *RSC Adv.* **2013**, *3*, 24758–24775. [CrossRef]

20. Chen, C.; Conception, J.J.; Jurss, J.W.; Meyer, T.J. Single-Site, Catalytic Water Oxidation on Oxide Surfaces. *J. Am. Chem. Soc.* **2009**, *131*, 15580–15581. [CrossRef]

21. Xie, K.; Sun, L.; Wang, C.; Lai, Y.; Wang, M.; Chen, H.; Lin, C. Photocatalytic properties of Ag nanoparticles loaded TiO$_2$ nanotube arrays prepared by pulse current deposition. *Electrochim. Acta* **2010**, *55*, 7211–7218. [CrossRef]

22. Swarnakar, P.; Kanel, S.R.; Nepal, D.; Jiang, Y.; Jia, H.; Kerr, L.; Goltz, M.N.; Levy, J.; Rakovan, J. Silver deposited titanium dioxide thin film for photocatalysis of organic compounds using natural light. *Solar Energy* **2013**, *88*, 242–249. [CrossRef]

23. Cheng, B.; Le, Y.; Yu, J. Preparation and enhanced photocatalytic activity of Ag@TiO$_2$ core–shell nanocomposite nanowires. *J. Hazard. Mat.* **2010**, *177*, 971–977. [CrossRef] [PubMed]

24. Palomares, E.; Vilar, R.; Durrant, J.R. Heterogeneous colorimetric sensor for mercuric salts. *Chem. Commun.* **2004**, 362–363. [CrossRef]

25. Mandal, S.S.; Bhattacharyya, A.J. Electrochemical sensing and photocatalysis using Ag-TiO$_2$ microwires. *J. Chem. Sci.* **2012**, *124*, 969–978. [CrossRef]

26. Jiang, Y.; Zheng, B.; Du, J.; Liu, G.; Guo, Y.; Xiao, D. Electrophoresis deposition of Ag nanoparticles on TiO$_2$ nanotube arrays electrode for hydrogen peroxide sensing. *Talanta* **2013**, *112*, 129–135. [CrossRef] [PubMed]

27. Ivanova, T.; Harizanova, A.; Koutzarova, T.; Vertuyen, B. Optical and structural characterization of TiO$_2$ films doped with silver nanoparticles obtained by a sol-gel method. *Opt. Mater.* **2013**, *36*, 207–213. [CrossRef]

28. Djaoued, Y.; Ozga, K.; Wojciechowski, A.; Reshak, A.H.; Robichaud, J.; Kityk, I.V. Photoinduced effects in TiO$_2$ nanocrystalline films with different morphology. *J. Alloys Compd.* **2010**, *508*, 599–605. [CrossRef]

29. Anandan, S.; Sathish Kumar, P.; Pugazhenthiran, N.; Madhavan, J.; Maruthamuthu, P. Effect of loaded silver nanoparticles for photocatalytic degradation of Acid Red 88. *Sol. Energy Mater. Sol. Cells* **2008**, *92*, 929–937. [CrossRef]

30. Stathatos, E.; Lianos, P.; Falaras, P.; Siokou, A. Photocatalytically deposited silver nanoparticles on mesoporous TiO$_2$ films. *Langmuir* **2000**, *16*, 2398–2400. [CrossRef]

31. Tricot, F.; Vocanson, F.; Chaussy, D.; Beneventi, D.; Reynaud, S.; Lefkir, Y.; Destouches, N. Photochromic Ag:TiO$_2$ thin films on PET substrate. *RSC Adv.* **2014**, *4*, 61305–61312. [CrossRef]

32. Okumu, J.; Dahmen, C.; Sprafke, A.N.; Luysberg, M.; von Plessen, G.; Wuttig, M. Photochromic silver nanoparticles fabricated by sputter deposition. *J. Appl. Phys.* **2005**, *97*, 094305. [CrossRef]

33. Ohko, Y.; Tatsuma, T.; Fujii, T.; Naoi, K.; Niwa, C.; Kubota, Y.; Fujishima, A. Multicolour photochromism of TiO 2 films loaded with silver nanoparticles. *Nat. Mater.* **2003**, *2*, 29–31. [CrossRef]

34. Hou, W.; Cronin, S.B. A review of surface plasmon resonance-enhanced photocatalysis. *Adv. Funct. Mater.* **2013**, *23*, 1612–1619. [CrossRef]

35. Wu, C.; Mosher, B.P.; Lyons, K.; Zeng, T. Reducing Ability and Mechanism for Polyvinylpyrrolidone (PVP) in Silver Nanoparticles Synthesis. *J. Nanosci. Nanotechnol.* **2010**, *10*, 2342–2347. [CrossRef]

36. Zhang, S.; Tang, Y.; Vlahovic, B. Preparation of Silver Nanoparticles in Poly(N-vinylpyrrolidone)/Ethanol Solutions. *Int. J. Nanosci. Ser.* **2017**, *16*, 1750008. [CrossRef]

37. Mierzwa, M.; Lamouroux, E.; Walcarius, A.; Etiene, M. Porous and transparent metal-oxide electrodes: Preparation methods and electroanalytical application prospects. *Electroanalysis* **2018**, *30*, 1241–1258. [CrossRef]

38. Hirakawa, T.; Kamat, P.V. Charge separation and catalytic activity of Ag@TiO$_2$ core-shell composite clusters under UV-irradiation. *J. Am. Chem. Soc.* **2005**, *127*, 3928–3934. [CrossRef]

39. Nadar, L.; Sayah, R.; Vocanson, F.; Crespo-Montteiro, N.; Boukenter, A.; Joao, S.S.; Destouches, N. Influence of reduction process of the colour and photochromism of amorphous mesoporous TiO$_2$ thin films loaded with a silver salt. *Photochem. Photobiol. Sci.* **2011**, *10*, 1810–1816. [CrossRef]

40. Kawahara, K.; Suzuki, K.; Okho, Y.; Tatsuma, T. Electron transport in silver-semiconductor nanocomposite films exhibiting multicolor photochromism. *Phys. Chem. Chem. Phys.* **2005**, *7*, 3851–3855. [CrossRef]

41. Awazu, K.; Fujimaki, M.; Rockstuhl, C.; Tominaga, J.; Murakami, H.; Ohki, Y.; Yoshida, N.; Wanatabe, T. A plasmonic photocatalyst consisting of silver nanoparticles embedded in titanium dioxide. *J. Am. Chem. Soc.* **2008**, *130*, 1676–1680. [CrossRef] [PubMed]

42. He, J.; Ichinose, I.; Kunitake, T.; Nakao, A. In situ synthesis of noble metal nanoparticles in ultrathin TiO$_2$-Gel films by a combination of on ion-exchange and reduction processes. *Langmuir* **2002**, *18*, 10005–10010. [CrossRef]

43. Gharibshahi, L.; Saion, E.; Gharibshahi, E.; Shaari, A.H.; Matori, K.A. Structural and Optical Properties of Ag Nanoparticles Synthesized by Thermal Treatment Method. *Materials* **2017**, *10*, 402. [CrossRef]

44. Sun, Y.; Xia, Y. Triangular Nanoplates of Silver: Synthesis, Characterization, and Use as Sacrificial Templates for Generating Triangular Nanorings of Gold. *Adv. Mater.* **2003**, *15*, 695–698. [CrossRef]

45. Arabatzis, I.M.; Stergiopoulos, T.; Bernard, M.C.; Labou, D.; Neophytides, S.G.; Falaras, P. Silver-modified titanium dioxide thin films for efficient photodegradation of methyl orange. *Appl. Catal. B Environ.* **2003**, *42*, 187–201. [CrossRef]

46. Couzon, N.; Maillard, M.; Bois, L.; Chassagneux, F.; Brioude, A. Electrochemical observation of the plasmonic effect in photochromic Ag nanoparticle filled mesoporous TiO$_2$ film. *J. Phys. Chem. C* **2017**, *121*, 22147–22155. [CrossRef]

47. Yu, B.; Leung, K.M.; Guo, Q.; Lau, W.M.; Yang, J. Synthesis of Ag-TiO$_2$ composite nano thin film for antimicrobial application. *Nanotechnology* **2011**, *22*, 115603. [CrossRef]

48. Kuzma, A.; Weis, M.; Flickyngerova, S.; Jakabovic, J.; Satka, A.; Dobrocka, E.; Chlpik, J.; Cirak, J.; Donoval, M.; Telek, P.; et al. Influence of surface oxidation on plasmon resonance in monolayer of gold and silver nanoparticles. *J. Appl. Phys.* **2012**, *112*, 103531. [CrossRef]

49. Kim, Y.S.; Balland, V.; Limoges, B.; Costentin, C. Cyclic Voltammetry modelling of proton transport effects on redox charge storage in conductive materials: Application to a mesoporous TiO_2 film. *Phys. Chem. Chem. Phys.* **2017**, *19*, 17944–17951. [CrossRef]

50. Fabregat-Santiago, F.; Randriamahazaka, H.; Zaban, A.; Garcia-Canadas, J.; Garcia-Belmonte, G.; Bisquert, J. Chemical capacitance of nanoporous-nanocrystalline TiO_2 in a room temperature ionic liquid. *Phys. Chem. Chem. Phys.* **2006**, *8*, 1827–1833. [CrossRef]

51. Fabregat-Santiago, F.; Garcia-Belmonte, G.; Bisquert, J.; Bogdanoff, P.; Zaban, A. Mott-Schottky analysis of nanoporous semiconductor electrodes in dielectric state deposited on SnO_2 (F) conducting substrates. *J. Electrochem. Soc.* **2002**, *150*, E293. [CrossRef]

52. Lyon, L.A.; Hupp, J.T. Energetics on the nanocrystalline titanium dioxide/aqueous solution interface: Approximate conduction band edge variations between $H_0 = -10$ and $H- = +26$. *J. Phys. Chem. B* **1999**, *103*, 4623–4628. [CrossRef]

53. Cuteo-Gomez, L.F.; Garcia-Gomez, N.A.; Mosqueda, H.A.; Sanchez, E.M. Electrochemical study of TiO_2 modified with silver nanoparticles upon CO_2 reduction. *J. Appl. Electrochem.* **2014**, *44*, 675–682. [CrossRef]

54. Renault, C.; Nicole, L.; Sanchez, C.; Costentin, C.; Balland, V.; Limoges, B. Unraveling the charge transfer/electron transport in mesoporous semiconductive TiO_2 films by voltabsoptometry. *Phys. Chem. Chem. Phys.* **2015**, *17*, 10592–10607. [CrossRef]

55. Wang, G.F.; Li, M.G.; Gao, Y.C.; Fang, B. Amperometric sensor used for determination of thiocyanate with a silver nanoparticles modified electrode. *Sensors* **2004**, *4*, 147–155. [CrossRef]

56. Chang, G.; Zhang, J.; Oyama, M.; Hirao, K. Silver-nanoparticle-attached indium tin oxide surfaces fabricated by a seed-mediated growth approach. *J. Phys. Chem. B* **2005**, *109*, 1204–1209. [CrossRef]

Article

A Rapid Synthesis of Mesoporous Mn_2O_3 Nanoparticles for Supercapacitor Applications

You-Hyun Son [1], Phuong T. M. Bui [1], Ha-Ryeon Lee [1], Mohammad Shaheer Akhtar [1,2,*], Deb Kumar Shah [1] and O-Bong Yang [1,2,*]

[1] School of Semiconductor and Chemical Engineering & Solar Energy Research Center, Chonbuk National University, Jeonju 54896, Korea; youheun94@jbnu.ac.kr (Y.-H.S.); missbui0504@gmail.com (P.T.M.B.); haryeonlee@jbnu.ac.kr (H.-R.L.); dkshah149@jbnu.ac.kr (D.K.S.)

[2] New and Renewable Energy Materials Development Center (NewREC), Chonbuk National University, Jeonbuk 54896, Korea

* Correspondence: shaheerakhtar@jbnu.ac.kr (M.S.A.); obyang@jbnu.ac.kr (O-B.Y.)

Received: 20 August 2019; Accepted: 24 September 2019; Published: 30 September 2019

Abstract: Mn_2O_3 nanomaterials have been recently composing a variety of electrochemical systems like fuel cells, supercapacitors, etc., due to their high specific capacitance, low cost, abundance and environmentally benign nature. In this work, mesoporous Mn_2O_3 nanoparticles (NPs) were synthesized by manganese acetate, citric acid and sodium hydroxide through a hydrothermal process at 150 °C for 3 h. The synthesized mesoporous Mn_2O_3 NPs were thoroughly characterized in terms of their morphology, surfaces, as well as their crystalline, electrochemical and electrochemical properties. For supercapacitor applications, the synthesized mesoporous Mn_2O_3 NP-based electrode accomplished an excellent specific capacitance (C_{sp}) of 460 $F \cdot g^{-1}$ at 10 $mV \cdot s^{-1}$ with a good electrocatalytic activity by observing good electrochemical properties in a 6 M KOH electrolyte. The excellent C_{sp} might be explained by the improvement of the surface area, porous surface and uniformity, which might favor the generation of large active sites and a fast ionic transport over the good electrocatalytic surface of the Mn_2O_3 electrode. The fabricated supercapacitors exhibited a good cycling stability after 5000 cycles by maintaining ~83% of C_{sp}.

Keywords: Mn_2O_3; mesoporous materials; electrochemical characterizations; electrode; supercapacitors

---

1. Introduction

A popular electrochemical energy-storage system, the supercapacitor has been a well-explored device as a heartening energy storage because of its high-power density, marvelous cycling and fast charge-discharge mechanisms [1–4]. Supercapacitors, on the basis of their charge-storage process, are classified into two types: i) double layer electrochemical capacitors (EDLCs), built from the electrode and electrolyte interface for the ions adsorption-desorption, and ii) the electrode materials-governed faradaic reaction-based pseudocapacitors [5–7]. In comparison with the EDLCs, the pseudocapacitors have exhibited a fast and desirable reversible redox reaction that promotes an excellent charging and discharging process, resulting in the enhancement of the charge storage capacities. In addition, pseudocapacitors display a highly competent storage device for rechargeable batteries owing to their fast energy harvesting, high energy and high-power delivery [8–10]. In supercapacitors, the active electrodes are normally prepared with carbon-based materials, conducting polymers, and a variety of transition metal oxide materials [11–14]. In recent years, transition metal oxides, such as RuO_2, MnO_2, Mn_3O_4, NiO, Nb_2O_5, V_2O_5, CoO_x, MoO_3, and TiO_2, have been frequently applied in preparing an effective electrode for supercapacitors due to their abundance in nature, inexpensive nature and extraordinary redox activity [13–21].

Apart from other transition metal oxides, the manganese-based oxides (MnO_x) and their derivatives, like MnO_2, Mn_2O_3, and Mn_3O_4, are popularly used as electrode materials in supercapacitors because they have a non-toxic nature, good structural flexibility and excellent chemical and physical stability in various electrolytes [22]. In particular, Mn_2O_3 materials as anode materials in lithium ion batteries have shown a high capacity and demonstrate a theoretically high specific capacity of ~1018 mAh/g, while also exhibiting a high specific capacitance [23]. Until now, the Mn_2O_3 materials-based electrodes in supercapacitors have been less explored and could be expected to have a high capacity and storage properties as they show an excellent environment compatibility and a resistance in acidic/alkaline electrolytes. Numerous efforts have been made to improve the performance of Mn_2O_3-based electrodes by adopting various modifications, such as chemical modifications [24], the incorporation of high surface-area conductive materials [25,26] and nanostructure fabrication [24,27]. Therefore, the synthesis of Mn_2O_3 materials with a controlled size, morphology and density through a cost-effective, simple and environment friendly method could be potentially sound [28,29]. It is expected that mesoporous Mn_2O_3 nanomaterials might show a surface that is fruitful for fast ion transportation in electrochemical supercapacitors [30].

In this work, a rapid and low temperature hydrothermal process was used to synthesize well-crystalline mesoporous Mn_2O_3 materials that were successfully applied as electro-active materials for a pseudosupercapacitor. The prepared mesoporous Mn_2O_3 materials-based electrode exhibits a high specific capacitance of 460 $F \cdot g^{-1}$ at a scan rate of 10 $mV \cdot s^{-1}$, with a good cycling stability after 5000 cycles.

2. Materials and Methods

In a typical synthesis, a mixture of 1 g of manganese acetate ($Mn(CH_3CO_2)_2$, Sigma-Aldrich, Saint Louis, MO, USA) and 0.5 g citric acid (Samchun Chemicals, Seoul, Korea) was dissolved in 100 mL of deionized (DI) water. Using an aqueous 5 M NaOH solution, the pH of the reaction mixture was adjusted to pH ~10 as the solution color changed from bright brown to dark brown. The hydrothermal process was carried out by transferring the reaction mixture into a Teflon beaker that was tightly sealed. Finally, a temperature of 150 °C was maintained for 3 h. After cooling down the autoclave, the obtained precipitates were collected by filtration, washed several times with DI water and with ethanol. The collected precipitates were dried in an oven overnight at 80 °C, and finally the obtained black powder was calcined at 500 °C in 1 h to remove other impurities.

For the electrochemical supercapacitor application, the synthesized mesoporous Mn_2O_3 electrode was prepared by mixing 85 wt.% Mn_2O_3 powders, 10 wt.% carbon black (Super P, Venatech, Seoul, South Korea), 3 wt.% carboxyl methyl cellulose (CMC, Sigma-Aldrich) and 2 wt.% polytetrafluoroethylene (PTFE, TCI chemical, Tokyo, Japan) in DI water to obtain a paste that was spread over the Ni foam using a glass rod via the rolling method. Afterward, the mesoporous Mn_2O_3-coated Ni foam electrodes were dried in the oven at 80 °C for 20 min to remove the solvent. For the electrochemical measurement, a three-electrode system, comprised of Mn_2O_3-coated Ni foam as the working electrode, Pt wire as the counter electrode and Ag/AgCl as the reference electrode, was used, and a cyclicvoltametric measurement (VersaSTAT4, AMETEK, Inc., Berwyn, PA, USA) was performed in an aqueous 6 M KOH electrolyte. All cyclicvoltametry (CV) measurements were observed at different scan rates ranging from 10 to 500 $mV \cdot s^{-1}$ in the voltage range of 0 to 1.0 V. A potentiostat/galvanostat (VersaSTAT4, AMETEK, Inc.) was used to analyze the electrochemical impedance spectroscopy (EIS) of the fabricated supercapacitor, based on the mesoporous Mn_2O_3 electrode, with a frequency ranging from 0.1 Hz to 1 MHz. For the calculation of C_{sp}, the mass loading of the mesoporous Mn_2O_3 on the electrode was ~0.001 g.

3. Results and Discussion

The morphological features of the synthesized mesoporous Mn_2O_3 materials were analyzed by field emission scanning electron microscopy (FESEM, Hitachi S-4800, Tokyo, Japan) and transmission electron microscopy (TEM, JEM-ARM200F, JEOL, Peabody, MA, USA) observations. Figure 1a,b shows the FESEM images of the synthesized mesoporous Mn_2O_3 materials at low and high magnifications. The spherical small particles, which are highly uniform, are visible in Figure 1a. From FESEM observations, it is difficult to identify the porosity of the synthesized materials. At a high magnification mode (Figure 1b), the obtained particles possess highly porous structures with average sizes of 10–30 nm. A similar observation for synthesized mesoporous Mn_2O_3 materials was detected in the TEM and high-resolution transmission electron microscopy (HRTEM) analyses, as shown in Figure 1c,d. As seen in Figure 1c, the synthesized Mn_2O_3 materials show a similar spherical shape, with a few semi-spherical particles having average particle sizes of ~10–30 nm. A close look at Figure 1c shows that the porosity of the synthesized Mn_2O_3 materials might be defined by the presence of visible voids over the particle surfaces. As reported earlier [31], these voids in materials may be a detrimental factor of the porosity of the materials. The HRTEM image is shown in Figure 1d; it expresses clear lattice fringes from when the measurement was focused on one particle from Figure 1c. From the HRTEM image, the interplanar distance between two lattice fringes is estimated to be ~0.27 nm, which is well-indexed to the (222) plane of Mn_2O_3 [32].

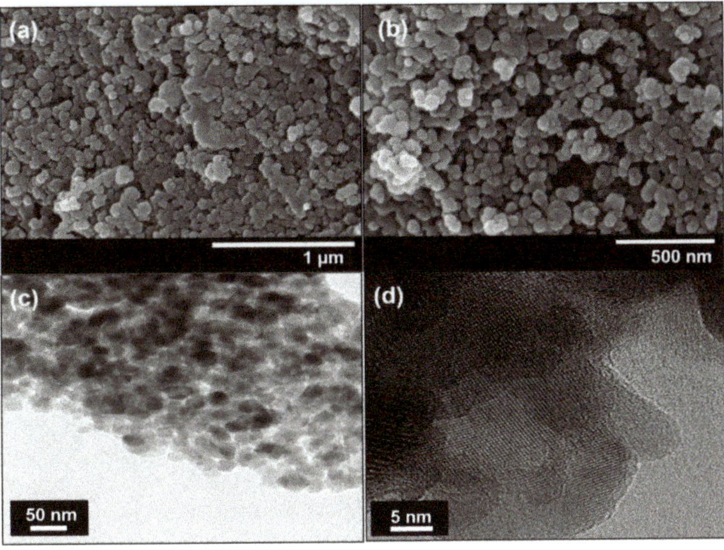

Figure 1. The (**a,b**) FESEM, (**c**) TEM and (**d**) HRTEM images of the synthesized mesoporous Mn_2O_3 materials.

The crystalline behavior and crystal planes of the synthesized mesoporous Mn_2O_3 materials were determined via a wide-angle X-ray diffraction (XRD, PANalytical, Malvern, United Kingdom) measurement, as shown in Figure 2. The well-defined diffraction peaks at 18.9°, 23.1°, 33.1°, 38.2°, 45.2°, 47.3°, 49.5°, 55.1° and 65.8° are associated to (200), (211), (222), (400), (332), (422), (431), (440) and (622) planes, respectively. All of the obtained diffraction peaks are assigned perfectly to the Bixbyite crystal phase α-Mn_2O_3, with JCPDS no. 41-1442 and space group Ia3, lattice constants $a = b = c = 9.4091$ Å, $\alpha = \beta = \gamma = 90°$. To estimate the crystallite sizes (CS) of the synthesized mesoporous Mn_2O_3 materials, the Debey–Scherrer equation was used [33]:

$$CS = \frac{0.95 \times \lambda}{\beta \cos(\theta)} \tag{1}$$

where β is the breadth of the observed diffraction line at its half-intensity maximum, K is the so-called shape factor, which usually takes a value of about 0.9, and λ is the wavelength of the X-ray source used in the XRD. By taking the maximum diffraction peak of (222), the crystallite size of the mesoporous Mn_2O_3 materials is found to be 28 nm, which is very close to the FESEM and TEM results.

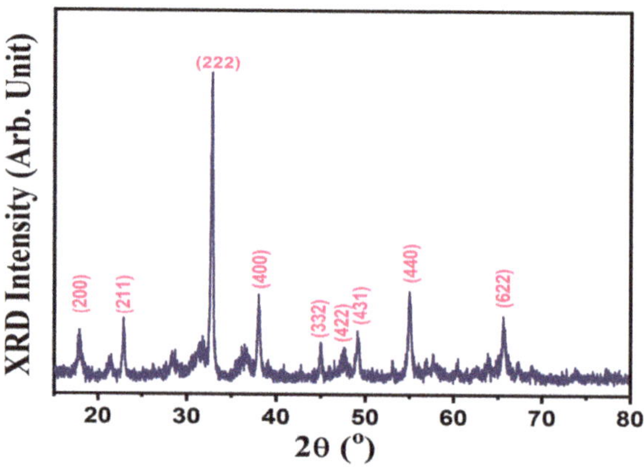

Figure 2. The XRD patterns of the synthesized mesoporous Mn_2O_3 materials.

Figure 3 shows the infrared (IR, Nicolet, IR300, Thermo Fisher Scientific, Waltham, MA, USA) and Raman spectroscopic studies (Raman microscope, Renishaw, UK) that define the structural properties of the synthesized mesoporous Mn_2O_3 materials. As seen in Figure 3a, two sharp IR bands are observed at 610 and at 520 cm^{-1}, assigning the stretching vibrations of Mn–O units and the asymmetric Mn–O–Mn stretching vibration, respectively [34]. Other IR bands at 1640 and 3343 cm^{-1} are detected, related to –OH and the water species from atmospheric moisture. It is believed that the observation of the IR bands at 520 and 610 cm^{-1} clearly reveals the formation of Mn_2O_3 without other impurities. Figure 3b depicts the Raman scattering spectroscopy of the synthesized mesoporous Mn_2O_3 materials. The mesoporous Mn_2O_3 NPs present a strong Raman band at 651 cm^{-1}, including with two weak Raman bands at 268 and at 175.0 cm^{-1}. The strong Raman band at 651 cm^{-1} represents the characteristic of the Mn_2O_3 along with the space group Ia3 structure [35], suggesting the typical symmetric stretching Mn–O–Mn bridge in Mn_2O_3. The main Raman band is well-matched with the reported literature of Mn_2O_3 [35]. Additionally, two weak Raman bands at ~268 and ~175 are assigned to the out-of-plane bending modes of Mn_2O_3 and the asymmetric stretching of the bridge oxygen species (Mn–O–Mn) [36], respectively.

To explain the thermal and structural properties, a thermal gravimetric analysis (TGA, Thermal analyzer, TA Instrument Ltd., New Castle, DE, USA) was performed for the synthesized mesoporous Mn_2O_3 materials, as displayed in Figure 4a,b. As seen in Figure 4b, four decomposition temperatures were visibly identified in the range of 25 to 800 °C. After 500 °C, the subsequent weight loss of ~1.5% that started from 500 to 600 °C is ascribed to the thermal decomposition of Mn_2O_3 to MnO, a decomposition that is usual in metal oxides. In the beginning, the synthesized mesoporous Mn_2O_3 materials were recorded as having a very small weight loss of ~0.5% up to 500 °C, which usually occurs via the removal of water/moisture from the sample. This suggests that the synthesized mesoporous Mn_2O_3 NPs exhibit a remarkably good stability with a high crystalline nature of Mn_2O_3.

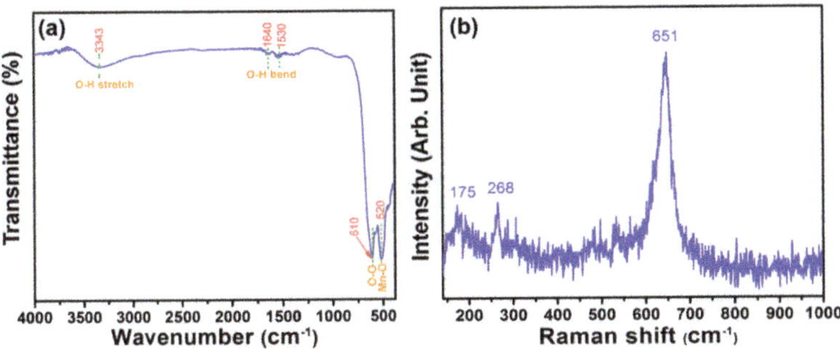

Figure 3. The (a) Fourier-transform infrared spectroscopy (FTIR) and (b) Raman spectrum of the synthesized mesoporous Mn_2O_3 materials.

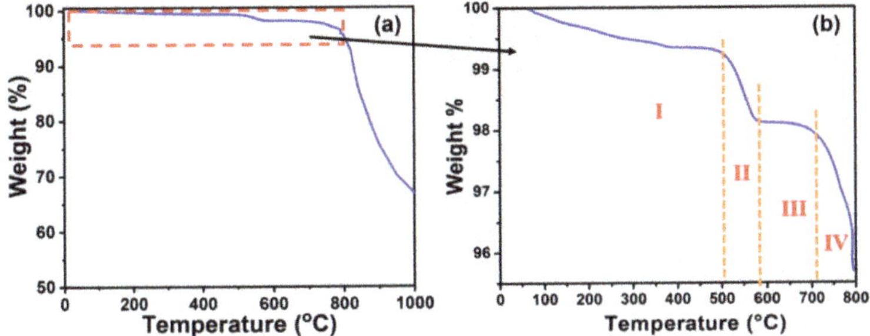

Figure 4. (a,b) TGA plot of the synthesized mesoporous Mn_2O_3 materials.

The synthesized mesoporous Mn_2O_3 materials were further characterized in terms of their composition and the oxidation states of the elements using an X-ray photoelectron spectroscopy (XPS, AXISNOVA CJ109, Kratos Inc., Manchester, UK) analysis. Figure 5a shows the survey XPS spectrum of the synthesized mesoporous Mn_2O_3 materials, revealing Mn 2p, Mn 3s and O 1s with weak C 1s peaks. The high-resolution Mn 2p XPS spectra of the synthesized mesoporous Mn_2O_3 materials is shown in Figure 5b and demonstrates doublet binding energies at 641.0 (Mn 2$p_{3/2}$) and 652.8 eV (Mn 2$p_{1/2}$). It is noted that the doublet Mn 2p, with an estimated spin–orbit splitting value of 11.8 eV, is nearly the same as the values reported for Mn_2O_3 [37]. Additionally, Figure 5c displays the high-resolution Mn 3s with characteristic doublet binding energies at 88.8 and 83.5 eV for Mn_2O_3. From the Mn 3s XPS, the peak separation for the doublet binding energies are ~5.3 eV, which is very close to the peak separation of Mn 3s in Mn_2O_3 [38], which again implies the formation of Mn_2O_3. Figure 5d depicts the high-resolution O 1s XPS spectrum. The two binding energies at 528.9 and 530.7 eV are normally related to oxygen O^{2-} in the lattice of Mn–O–Mn, indicating the formation of Mn_2O_3. Therefore, the XPS analysis implied the formation of a pure Mn_2O_3 form without any other oxide impurities [39].

The surface and porous behavior of the synthesized mesoporous Mn_2O_3 materials have been elucidated by analyzing the nitrogen (N_2) adsorption-desorption isotherms and BET surface analyzer, as shown Figure 6. The N_2 adsorption-desorption isotherms (Figure 6a) present a regular type IV isotherm with a small hysteresis in the relative pressure (p/p_0) range of 0.75–0.9, which clearly imitated the mesoporous characteristic of materials. Figure 6b shows how the pore size distribution (d) of the mesoporous Mn_2O_3 materials was recorded in the range of 10–30 nm, which is the case for mesoporous materials. Using a Brunauer-Emmett-Teller (BET) surface analysis, the specific surface area (s) for the synthesized mesoporous Mn_2O_3 materials was determined to be 76.9 $m^2 \cdot g^{-1}$. Thus, the synthesized Mn_2O_3 materials are mesoporous in nature which could provide a larger surface area for the high ion diffusion of ions in the electrolyte for high-performance supercapacitors.

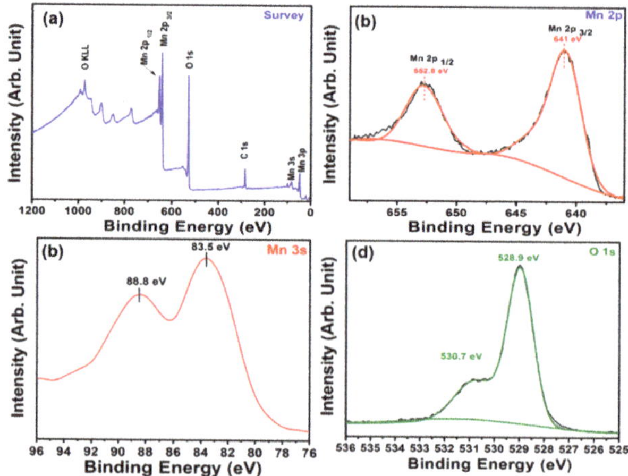

Figure 5. (**a**) Survey, (**b**) high resolution Mn 2*p*, (**c**) Mn 3*s*, and (**d**) O 1*s* XPS of the synthesized mesoporous Mn$_2$O$_3$ materials.

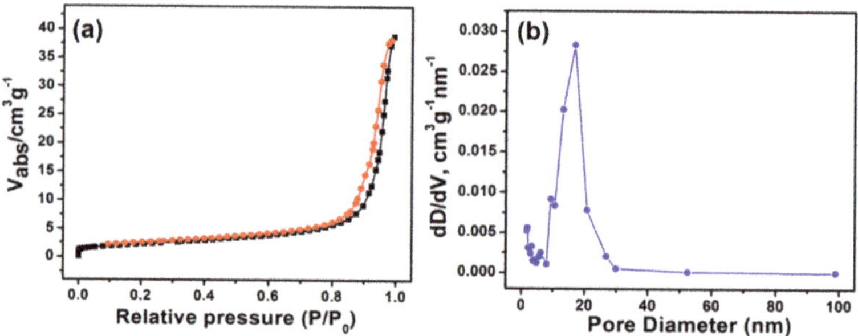

Figure 6. The (**a**) N$_2$ adsorption-desorption isotherm and (**b**) pore size distribution plot of the synthesized mesoporous Mn$_2$O$_3$ materials.

The synthesized mesoporous Mn$_2$O$_3$ materials were utilized as electroactive materials to evaluate the supercapacitor properties. The parameters for the supercapacitor with the synthesized mesoporous Mn$_2$O$_3$ electrode were determined by measuring the cyclicvoltametry (CV) at different scan rates in a 6 M KOH electrolyte. Figure 7a shows a series of CV curves of the synthesized mesoporous Mn$_2$O$_3$ electrode at different scan rates in a 6 M KOH electrolyte. Generally, in an electrochemical reaction, the conversion from Mn^{3+} to Mn^{4+} in an Mn$_2$O$_3$ electrode occurs via an oxidation reaction. This oxidation reaction normally accelerates the reaction kinetics of OH$^-$ over the Mn$_2$O$_3$ cubic lattice through the chemisorbed and/or intercalated. The charge/discharge process can be explained by the following electrochemical reaction, which involves the chemisorption/intercalation of HO$^-$ over the Mn$_2$O$_3$ surfaces:

$$Mn_2O_3 + HO^- \underset{Dischage}{\overset{Charge}{\longleftrightarrow}} Mn^{4+}[Mn^{3+}]OH^-O_3 + \bar{e}$$

The prepared Mn$_2$O$_3$ electrode depicts the oxidation-reduction pair peaks with different curve shapes of the charge–discharge curves in relation to the variation in the scan rates. The observations principally suggest the origin of the faradaic pseudo-capacitance in an alkaline electrolyte over the surface of an Mn$_2$O$_3$ electrode. Moreover, the well-defined oxidation reduction pair within 0–1 V can also be ascribed to a faradaic redox reaction [40,41]. The plot of the capacitance versus the scan rate for a fabricated pseudosupercapacitor based on a mesoporous Mn_2O_3 electrode is displayed in Figure 7b. In general, the specific capacitance (C_{sp}) is estimated from the CV curves using the following equation [42]:

$$C_{sp} = \frac{1}{mv(V_2 - V_1)} \int_{V_1}^{V_2} I(V)dV \qquad (2)$$

where C_{sp} is the specific capacitance ($F \cdot g^{-1}$), I is the current (A), v is the scan rate, m is the mass of the active material (g) and ΔV is the potential range (V). As seen in Figure 7b, the fabricated pseudosupercapacitor based on the mesoporous Mn_2O_3 electrode exhibits a high C_{sp} of ~460 $F \cdot g^{-1}$ at a scan rate of 10 $mV \cdot s^{-1}$. This might be explained by the fact that there is an improvement of the surface area, the porous surface and uniformity, which might favor the generation of large active sites [43,44] and fast ionic transport over the surface of the Mn_2O_3 electrode. Specifically, the porous structure could provide a large and accessible surface area to ion adsorption, improve the accessibility of cations and shorten the ion diffusion path. The stability of the mesoporous Mn_2O_3 electrode is explored by observing the multicycle CV measurements in a 6 M KOH electrolyte. From Figure 7c, the slight shift in the oxidation peaks expresses the good stability of the mesoporous Mn_2O_3 electrode in the alkaline electrolyte. It is also seen that the anodic current in the Mn_2O_3 electrode is positively shifted with the increase, while the cathodic current is negatively shifted due to the increment in the electrical polarization and the fast and irreversible reactions as the scan rates increase. Here, the fast, irreversible reactions after 25 cycles might result from the accumulation of Mn^{4+} ions in the electrochemical system. The stability of the fabricated pseudosupercapacitor based on a mesoporous Mn_2O_3 electrode is shown in Figure 7d by measuring the capacitance after 5000 cycles. From Figure 7d, after 5000 cycles, the electrochemical system shows a reasonably high stability by retaining 83% of the initial capacity. Furthermore, the inset of Figure 7d presents the FESEM image of the mesoporous Mn_2O_3 electrode after 5000 cycles. The morphology of the Mn_2O_3 materials is not altered, except for an aggregation of small particles after cycling, suggesting the stability of the Mn_2O_3 electrode in the alkaline electrolyte. Hence, the reproducibility of a pseudosupercapacitor based on a mesoporous Mn_2O_3 electrode in a KOH electrolyte is remarkable and implies the low dissolution of the electro-active materials in a strong alkaline electrolyte after an electrochemical process.

The electrochemical Impedance Spectroscopy (EIS) for the fabricated pseudosupercapacitor based on the synthesized mesoporous Mn_2O_3 electrode was conducted to understand the electrochemical and electrical properties. Figure 8 shows the EIS plot, which was measured in the frequency range of $0.1–10^5$ Hz. In the illustration of the equivalent circuit, the starting point in the Nyquist plot at a high-frequency region and the starting point at a low-frequency region represent the series resistance (R_s) and the charge transfer resistance (R_{ct}) of the electrode/electrolyte interface, respectively. C_{dl}, W and C_{pseudo} explain the double layer capacitance that arose by the parallel connection to R_{ct}, the Warburg diffusion element and the Faradaic capacitance generated by the contribution of the Mn_2O_3 electrode. As shown in Figure 8, the pseudosupercapacitor based on the synthesized mesoporous Mn_2O_3 electrode features a large phase angle near the low-frequency region, indicating the faradic capacitance behavior of the electrode. Importantly, in the EIS plot, the straight line in the low-frequency region, and the absence of any small semicircle in the high-frequency region, are indicative of the good capacitive nature of the present electrochemical system based on a mesoporous Mn_2O_3 electrode. Likewise, the observed straight line at a low frequency might reduce the diffusion length and accelerate the ions transportation on the mesoporous surface of the Mn_2O_3 electrode, as evidenced by the CV results.

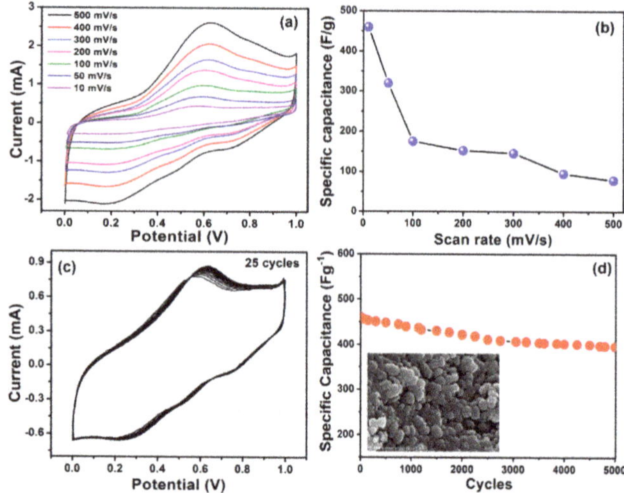

Figure 7. The (**a**) CV curves and (**b**) specific capacitance of the synthesized mesoporous Mn_2O_3 electrode at different scan rates ranging from 10 to 500 $mV \cdot s^{-1}$, (**c**) multicycles CV curves and (**d**) variation in the specific capacitance of the synthesized mesoporous Mn_2O_3 electrode after 5000 cycles. The inset shows the FESEM image of the synthesized mesoporous Mn_2O_3 electrode after 5000 cycles.

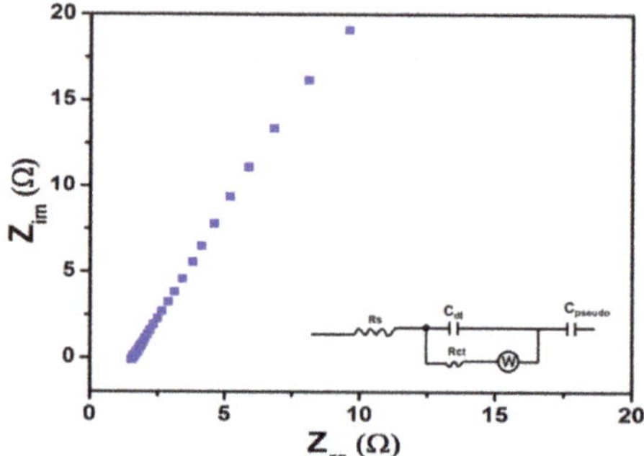

Figure 8. The EIS plot of the synthesized mesoporous Mn_2O_3 electrode at the Z_{im} versus the Z_{re} mode. The inset shows its corresponding equivalent circuit.

4. Conclusions

In summary, a facial hydrothermal process was adopted to synthesize well-crystalline mesoporous Mn_2O_3 materials for the fabrication of a pseudocapacitor. The crystalline and structural characterizations confirmed the formation of Mn_2O_3 materials without displaying any other oxide forms. The surface properties were analyzed, showing the mesoporous nature of Mn_2O_3 materials and an estimated high surface area of 76.9 $m^2 \cdot g^{-1}$ with a good pore distribution. The fabricated pseudo-capacitor based on a Mn_2O_3 mesoporous particle electrode shows a reasonably high specific capacitance of ~460 $F \cdot g^{-1}$ at 10 $mV \cdot s^{-1}$ in a 6 M KOH aqueous solution. The enhancement in the capacitive properties might be attributed to the high surface area, porous surface and uniformity of the unique mesoporous particles morphology, resulting in the large generation of active sites and a fast ionic transport over the surface of the Mn_2O_3 electrode.

Author Contributions: Conceptualization, Y.-H.S., P.T.M.B., M.S.A., H.-R.L., D.K.S. and O-B.Y.; methodology, Y.-H.S., P.T.M.B., M.S.A., H.-R.L. and D.K.S.; software, Y.-H.S. and P.T.M.B.; validation, P.T.M.B., M.S.A., H.-R.L. and D.K.S.; formal analysis, Y.-H.S. and P.T.M.B.; investigation, Y.-H.S. and P.T.M.B.; resources, Y.-H.S. and P.T.M.B.;

data curation, Y.-H.S., P.T.M.B., H.-R.L. and D.K.S.; writing—original draft preparation, Y.-H.S., P.T.M.B., and M.S.A.; writing—review and editing, Y.-H.S., P.T.M.B., M.S.A., and O-B.Y.; supervision, M.S.A., and O-B.Y.

Funding: This work was supported by the National Research Foundation of Korea (NRF) grant funded by the Korea government (MSIT) (NRF-2019R1F1A1063999).

Acknowledgments: This paper was also acknowledged the support of research funds from Chonbuk National University in 2018.

Conflicts of Interest: The authors declare no conflict of interest.

References

1. Miller, J.R.; Simon, P. Electrochemical capacitors for energy management. *Science* **2008**, *321*, 651–652. [CrossRef] [PubMed]
2. Simon, P.; Gogotsi, Y. Materials for electrochemical capacitors. *Nat. Mater.* **2008**, *7*, 845–854. [CrossRef]
3. Wei, D.; Scherer, M.R.; Bower, C.; Andrew, P.; Ryhanen, T.; Steiner, U. A nanostructured electrochromic supercapacitor. *Nano Lett.* **2012**, *12*, 1857–1862. [CrossRef] [PubMed]
4. Yuan, C.Z.; Wu, H.B.; Xie, Y.; Lou, X.W. Mixed transition-metal oxides: Design, synthesis, and energy-related applications. *Angew. Chem. Int. Ed.* **2014**, *53*, 1488–1504. [CrossRef] [PubMed]
5. Jiang, H.; Li, C.; Ma, J. Polyaniline–MnO_2 coaxial nanofiber with hierarchical structure for high-performance supercapacitors. *J. Mater. Chem.* **2012**, *22*, 2751. [CrossRef]
6. Devaraj, S.; Munichandraiah, N. High capacitance of electrodeposited MnO_2 by the effect of a surface-active agent. *Electrochem. Solid State Lett.* **2005**, *8*, 373–377. [CrossRef]
7. Zhao, D.; Tian, J.S.; Ji, Q.Q.; Zhang, J.T.; Zhao, X.S.; Guo, P.Z. Mn_2O_3 nanomaterials: Facile synthesis and electrochemical properties. *Chin. J. Inorg. Chem.* **2010**, *26*, 832–838.
8. Jang, G.; Ameen, S.; Akhtar, M.S.; Kim, E.-B.; Shin, H.S. Electrochemical Investigations of Hydrothermally Synthesized Porous Cobalt Oxide (Co3O4) Nanorods: Supercapacitor Application. *ChemistrySelect* **2017**, *2*, 8941–8949. [CrossRef]
9. Li, Z.Y.; Bui, P.T.M.; Kwak, D.H.; Akhtar, M.S.; Yang, O.B. Enhanced electrochemical activity of low temperature solution process synthesized Co_3O_4 nanoparticles for pseudo-supercapacitors applications. *Ceram. Int.* **2016**, *42*, 1879–1885. [CrossRef]
10. Li, Z.Y.; Akhtar, M.S.; Yang, O.B. Supercapacitors with ultrahigh energy density based on mesoporous carbon nanofibers: Enhanced double-layer electrochemical properties. *J. Alloy. Compd.* **2015**, *653*, 212–218. [CrossRef]
11. Seo, M.K.; Saouab, A.; Park, S.J. Effect of annealing temperature on electrochemical characteristics of ruthenium oxide/multi-walled carbon nanotube composites. *Mater. Sci. Eng. B* **2010**, *167*, 65–69. [CrossRef]
12. Shambharkar, B.H.; Umare, S.S. Production and characterization of polyaniline/Co_3O_4 nanocomposite as a cathode of Zn–polyaniline battery. *Mater. Sci. Eng. B* **2010**, *175*, 120–128. [CrossRef]
13. Zheng, Z.; Huang, L.; Zhou, Y.; Hu, X.W.; Ni, X.M. Large-scale synthesis of mesoporous CoO-doped NiO hexagonal nanoplatelets with improved electrochemical performance. *Solid State Sci.* **2009**, *11*, 1439–1443. [CrossRef]
14. Prasad, K.R.; Miura, N. Electrochemical synthesis and characterization of nanostructured tin oxide for electrochemical redox supercapacitors. *Electrochem. Commun.* **2004**, *6*, 849–852. [CrossRef]
15. Jiang, H.; Zhao, T.; Ma, J.; Yan, C.; Li, C. Ultrafine manganese dioxide nanowire network for high-performance supercapacitors. *Chem. Commun.* **2011**, *47*, 1264. [CrossRef] [PubMed]
16. Rakhi, R.B.; Chen, W.; Cha, D.; Alshareef, H.N. Substrate dependent self-organization of mesoporous cobalt oxide nanowires with remarkable pseudocapacitance. *Nano Lett.* **2012**, *12*, 2559. [CrossRef] [PubMed]
17. Lu, Q.; Lattanzi, M.W.; Chen, Y.; Kou, X.; Li, W.; Fan, X.; Unruh, K.M.; Chen, J.G.; Xiao, J.Q. Supercapacitor electrodes with high-energy and power densities prepared from monolithic NiO/Ni nanocomposites. *Angew. Chem.* **2011**, *50*, 6847–6850. [CrossRef]
18. Benson, J.; Boukhalfa, S.; Magasinski, A.; Kvit, A.; Yushin, G. Chemical vapor deposition of aluminum nanowires on metal substrates for electrical energy storage applications. *ACS Nano* **2012**, *6*, 118–125. [CrossRef]
19. Xie, K.; Li, J.; Lai, Y.; Lu, W.; Zhang, Z.; Liu, Y.; Zhou, L.; Huang, H. Highly ordered iron oxide nanotube arrays as electrodes for electrochemical energy storage. *Electrochem. Commun.* **2011**, *13*, 657–660. [CrossRef]

20. Ghosh, S.; Polaki, S.R.; Sahoob, G.; Jin, E.-M.; Kamruddin, M.; Cho, J.S.; Jeong, S.M. Designing metal oxide-vertical graphene nanosheets structures for 2.6 V aqueous asymmetric electrochemical capacitor. *J. Ind. Eng. Chem.* **2019**, *72*, 107–116. [CrossRef]

21. Nagamuthu, S.; Ryu, K.-S. MOF-derived microstructural interconnected network porous Mn_2O_3/C as negative electrode material for asymmetric supercapacitor device. *CrystEngComm* **2019**, *21*, 1442–1451. [CrossRef]

22. Sui, N.; Duan, Y.; Jiao, X.; Chen, D. Large-scale preparation and catalytic properties of one-dimensional α/β-MnO_2 nanostructures. *J. Phys. Chem. C* **2009**, *113*, 8560–8565. [CrossRef]

23. Kolathodi, M.S.; Rao, S.N.H.; Natarajana, T.S.; Singh, G. Beaded manganese oxide (Mn_2O_3) nanofibers: Preparation and application for capacitive energy storage. *J. Mater. Chem. A* **2016**, *4*, 7883–7891. [CrossRef]

24. Nakayama, M.; Tanaka, A.; Sato, Y.; Tonosaki, T.; Ogura, K. Electrodeposition of manganese and molybdenum mixed oxide thin films and their charge storage properties. *Langmuir* **2005**, *21*, 5907. [CrossRef]

25. Wu, Z.-S.; Ren, W.; Wang, D.-W.; Li, F.; Liu, B. High-energy MnO_2 nanowire/graphene and graphene asymmetric electrochemical capacitors. *ACS Nano* **2010**, *4*, 5835–5842. [CrossRef]

26. Fischer, A.E.; Pettigrew, K.A.; Rolison, D.R.; Stroud, R.M.; Long, J.W. Incorporation of homogeneous, nanoscale MnO_2 within ultraporous carbon structures via self-limiting electroless deposition: Implications for electrochemical capacitors. *Nano Lett.* **2007**, *7*, 281–286. [CrossRef]

27. Ghodbane, O.; Pascal, J.-L.; Favier, F. Microstructural effects on charge-storage properties in MnO_2-based electrochemical supercapacitors. *ACS Appl. Mater. Interfaces* **2009**, *1*, 1130–1139. [CrossRef]

28. Zhi, M.; Xiang, C.; Li, J.; Li, M.; Wu, N. Nanostructured carbon–metal oxide composite electrodes for supercapacitors: A review. *Nanoscale* **2013**, *5*, 72–88. [CrossRef]

29. Jiang, J.; Li, Y.; Liu, J.; Huang, X.; Yuan, C.; Lou, X.W. Recent advances in metal oxide-based electrode architecture design for electrochemical energy storage. *Adv. Mater.* **2012**, *24*, 5166–5180. [CrossRef]

30. Han, Z.J.; Seo, D.H.; Yick, S.; Chen, J.H.; Ostrikov, K. MnOx/carbon nanotube/reduced graphene oxide nanohybrids as high-performance supercapacitor electrodes. *NPG Asia Mater.* **2014**, *6*, e140. [CrossRef]

31. Demir, M.; Ashourirad, B.; Mugumya, J.H.; Saraswata, S.K.; El-Kaderi, H.M.; Gupta, R.B. Nitrogen and oxygen dual-doped porous carbons prepared from pea protein as electrode materials for high performance supercapacitors. *Int. J. Hydrog. Energy* **2018**, *43*, 18549–18558. [CrossRef]

32. Cheng, F.; Shen, J.; Ji, W.; Tao, Z.; Chen, J. Selective synthesis of manganese oxide nanostructures for electrocatalytic oxygen reduction. *ACS Appl. Mater. Interfaces* **2009**, *1*, 460–466. [CrossRef] [PubMed]

33. Jenkins, R. *Introduction to X-Ray Powder Diffractometry*; John Wiley & Sons, Inc.: New York, NY, USA, 1996; p. 90.

34. Gillot, B.; El Guendouzi, M.; Laarj, M. Particle size effects on the oxidation–reduction behavior of Mn_3O_4 hausmannite. *Mater. Chem. Phys.* **2001**, *70*, 54–60. [CrossRef]

35. Chen, Z.W.; Lai, J.K.L.; Shek, C.H. Influence of grain size on the vibrational properties in Mn_2O_3 nanocrystals. *J. Non Cryst. Solids* **2006**, *352*, 3285–3289. [CrossRef]

36. Han, Y.-F.; Ramesh, K.; Chen, L.; Widjaja, E.; Chilukoti, S.; Chen, F. Controlled synthesis, characterization, and catalytic properties of Mn2O3 and Mn3O4 nanoparticles supported on mesoporous silica SBA-15. *J. Phys. Chem. B* **2007**, *111*, 2830–2833.

37. Zhang, Y.; Yan, Y.; Wang, X.; Li, G.; Deng, D.; Jiang, L.; Shu, C.; Wang, C. Facile synthesis of porous Mn_2O_3 nanoplates and their electrochemical behavior as anode materials for lithium ion batteries. *Chem. A Eur. J.* **2014**, *20*, 6126–6130. [CrossRef]

38. Jahan, M.; Tominaka, S.; Henzie, J. Phase pure α-Mn_2O_3 prisms and their bifunctional electrocatalytic activity in oxygen evolution and reduction reactions. *Dalton Trans.* **2016**, *45*, 18494–18501. [CrossRef]

39. Ilton, E.S.; Post, J.E.; Heaney, P.J.; Ling, F.T.; Kerisit, S.N. XPS determination of Mn oxidation states in Mn (hydr)oxides. *Appl. Surf. Sci.* **2016**, *366*, 475–485. [CrossRef]

40. Toupin, M.; Brousse, T.; Bélanger, D. Influence of microstucture on the charge storage properties of chemically synthesized manganese dioxide. *Chem. Mater.* **2002**, *14*, 3946–3952. [CrossRef]

41. Toupin, M.; Brousse, T.; Bélanger, D. Charge storage mechanism of MnO_2 electrode used in aqueous electrochemical capacitor. *Chem. Mater.* **2004**, *16*, 3184–3190. [CrossRef]

42. Rajkumar, M.; Hsu, C.-T.; Wu, T.-H.; Chen, M.-G.; Hu, C.-C. Advanced materials for aqueous supercapacitors in the asymmetric design. *Prog. Nat. Sci. Mater. Int.* **2015**, *25*, 527. [CrossRef]

43. Gopalakrishnan, I.K.; Bagkar, N.; Ganguly, R.; Kulshreshtha, S.K. Synthesis of superparamagnetic Mn_3O_4 nanocrystallites by ultrasonic irradiation. *J. Cryst. Growth* **2005**, *280*, 436. [CrossRef]
44. Li, Z.-Y.; Akhtar, M.S.; Bui, P.T.M.; Yang, O.-B. Predominance of two dimensional (2D) Mn_2O_3 nanowalls thin film for high performance electrochemical supercapacitors. *Chem. Eng. J.* **2017**, *330*, 1240–1247. [CrossRef]

 *coatings*

Article

Increasing the Efficiency of Dye-Sensitized Solar Cells by Adding Nickel Oxide Nanoparticles to Titanium Dioxide Working Electrodes

Chih-Hung Tsai *, Chia-Ming Lin and Yen-Cheng Liu

Department of Opto-Electronic Engineering, National Dong Hwa University, Hualien 97401, Taiwan; 410125015@gms.ndhu.edu.tw (C.-M.L.); 410425072@gms.ndhu.edu.tw (Y.-C.L.)
* Correspondence: cht@mail.ndhu.edu.tw; Tel.: +886-3-890-3199

Received: 31 January 2020; Accepted: 21 February 2020; Published: 24 February 2020

Abstract: In this study, nickel oxide (NiO) nanoparticles were added to a titanium dioxide (TiO_2) nanoparticle paste to fabricate a dye-sensitized solar cell (DSSC) working electrode by using a screen-printing method. The effects of the NiO proportion in the TiO_2 paste on the TiO_2 working electrode, DSSC devices, and electron transport characteristics were comprehensively investigated. The results showed that adding NiO nanoparticles to the TiO_2 working electrode both inhibited electron transport (a negative effect) and prevented electron recombination with the electrolyte (a positive effect). The electron transit time was extended following an increase in the amount of NiO nanoparticles added, confirming that NiO inhibited electron transport. Furthermore, the energy level difference between TiO_2 and NiO generated a potential barrier that prevented the recombination of the electrons in the TiO_2 conduction band with the I_3^- ions in the electrolyte. When the TiO_2–NiO ratio was 99:1, the positive effects outweighed the negative effects. Therefore, this ratio was the optimal TiO_2–NiO ratio in the electrode for electron transport. The DSSCs with a TiO_2–NiO (99:1) working electrode exhibited an optimal power conversion efficiency of 8.39%, which was higher than the DSSCs with a TiO_2 working electrode.

Keywords: dye-sensitized solar cells; working electrode; TiO_2; NiO nanoparticles; electron transport

1. Introduction

The advancement of science and technology has increased the demand for energy over the years, resulting in a continuous reduction in oil reserves. Carbon dioxide produced by burning fossil fuels has caused global warming and has induced various anomalies in the global climate. Therefore, the development of green energy has attracted scientists' attention. Grätzel and colleagues introduced dye-sensitized solar cells (DSSCs) in 1991 [1]. Research related to solar cells immediately caught the public's attention because DSSCs have many advantages such as high efficiency, low cost, and simple fabrication [2–6]. Typical components of a DSSC include a titanium dioxide (TiO_2) working electrode, a dye, a platinum (Pt) counter electrode (CE), and an electrolyte [7–10]. The TiO_2 working electrode, which is used for transporting photoelectrons and exhibits a large surface area for dye adsorption and holes for injecting electrolytes, is a key component of a DSSC [11–13]. Although other types of wide bandgap oxides can achieve the same effects (e.g., ZnO [14], SnO_2 [15], Fe_2O_3 [16], and Nb_2O_5 [17]), TiO_2 is presently the optimal material for fabricating working electrodes. Previously, DSSCs contained planar TiO_2 working electrodes, which exhibit an efficiency of less than 1% because they rely on dye molecules adsorbed on the electrode surface for effective photocurrent generation [18]. Grätzel proposed porous electrodes comprising TiO_2 nanoparticles; each micrometer of thickness increases the surface area 100-fold, and the photocurrent generation efficiency can exceed 7% when dye molecules are adsorbed on the electrode surface. Accordingly, the fabrication of TiO_2, working electrodes is crucial in manufacturing high-efficiency DSSCs. Arakawa et al. designed six types of electrodes using three types of TiO_2 particles (23, 50, and 100 nm) and different mix proportions of TiO_2 particles, thereby revealing that a multilayered electrode achieved the optimal performance in photoelectron transport [19]. Kim et al. fabricated TiO_2 nanorods and mixed 10% of the nanorods with 90% of the TiO_2 nanoparticles; the authors reported that the photocurrent of DSSCs improved considerably when nanorods were added [20]. Recently, novel TiO_2 structures, such as nanowires and nanotubes, have been implemented to enhance the performance of TiO_2 working electrodes [21–24]. Various studies have been conducted to improve the design of TiO_2 working electrodes [25–27].

When light is irradiated on a DSSC, the dye converts from the ground state to the excited state and injects electrons into the conduction band of the TiO_2 working electrode, which then transports the electrons to the Pt counter electrode through the fluorine-doped tin oxide (FTO) conductive glass and external circuit. However, the transport does not thoroughly abide by scientific theory; reactions may occur during this procedure to affect the photocurrent and device efficiency. Studies have reported that the following cause a decreased photocurrent: (1) the reverse current caused by the reverse reaction between the dye and the electrolyte, (2) the recombination of electrons in the TiO_2 conduction band with the dye molecules, (3) the recombination of electrons in the TiO_2 conduction band with the electrolyte, and (4) the spontaneous recombination caused by a decline in the energy of the excited dye molecules [28]. Studies have indicated that barriers can be used to prevent the reverse current and recombination of electrons in the TiO_2 conduction band with the dye and electrolyte. Moreover, metal oxides can be applied as electron barriers to improve the transport efficiency of photoelectrons [29–31].

Nickel oxide (NiO), a transition metal oxide with a cubic lattice structure, can be synthesized through the solvothermal synthesis, precipitation calcination, chemical precipitation, microwave-assisted hydrothermal method, thermal decomposition, or sol–gel process. NiO is a p-type semiconductor material that is capable of electrocatalysis, high chemical stability, superconductivity, and electron transport. NiO has been widely researched and applied in catalysts, battery cathodes, gas detectors, electrochromic elements, magnetic materials, and photocathodes or counter electrodes in DSSCs [32–36]. Furthermore, the energy-level difference between p-type NiO and n-type TiO_2 may generate a potential barrier, which can prevent the recombination of the electrons in the TiO_2 conduction band with the I_3^- ions in the electrolyte, thus enhancing the efficiency of the DSSCs [37]. In this study, NiO nanoparticles were added to the TiO_2 nanoparticle solution to create a DSSC working electrode through screen printing, and the effects of the proportion of NiO in the TiO_2 solution on the TiO_2 working electrode, DSSC devices, and electron transport were investigated.

2. Experiments

2.1. Preparing the Materials

To prepare the ethyl cellulose (EC) solution, 5 g of EC (5–15 mPa·s, Sigma-Aldrich, St. Louis, MO, USA), 5 g of EC (30–60 mPa·s, Sigma-Aldrich), and 100 g of absolute alcohol (99.5%) were placed in a serum bottle, evenly mixed with a stir bar at 200 rpm, maintained at 40 °C for 3 days, and cooled to room temperature. To prepare the TiO_2 paste, 1.15 g of TiO_2 nanoparticles (particle size: 15–20 nm, Aeroxide TiO_2 P90), 3.4 g of terpineol (Merck, Darmstadt, Germany), and 4.5 g of the EC solution were placed in a container, combined with an appropriate amount of absolute alcohol, evenly mixed with a stir bar at 200 rpm, maintained at 40 °C for 3 days, and cooled to room temperature. To prepare the TiO_2–NiO combined paste, TiO_2 and NiO nanoparticles (particle size: 10–20 nm, US Research Nanomaterials, Inc., Houston, TX, USA) were added in proportions of 99:1 (1.1385 g, 0.0115 g), 98:2 (1.127 g, 0.023 g), and 97:3 (1.1155 g, 0.0345 g), respectively, to containers with 3.4 g of terpineol and 4.5 g of the EC solution. Subsequently, an appropriate amount of absolute alcohol was added to the solutions, which were then evenly mixed using stir bars at 200 rpm, maintained at 40 °C for 3 days, and later cooled to room temperature, thus yielding three TiO_2–NiO combined pastes in the following proportions: 99:1, 98:2, and 97:3. To prepare a 0.5-mM N719 dye solution, 11.87 mg of N719 dye, 3.92

mg of chenodeoxycholic acid (CDCA), 10 mL of acetonitrile (ACN), and 10 mL of tert-butanol were placed in a sample bottle and subjected to ultrasonic oscillation for 5 min.

2.2. Preparing the DCCS Devices

The FTO conductive glass substrate was cleaned and ultrasonically oscillated in acetone, deionized water, and alcohol (anhydrous ethanol, 99.5%, Echo Chemical Co., Ltd., Taiwan) for 5 min each and dried using nitrogen gas. The working electrode was then prepared by using the screen-printing method. The dried substrate was placed on the screen printer and the TiO_2 or TiO_2–NiO paste was printed evenly on the screen. A scraper was used to print from top to bottom so that the nanoparticle paste could pass through the screen and could reach the substrate. The printed substrate was placed in a culture dish, covered with another culture dish, sprayed with alcohol (anhydrous ethanol, 99.5%, Echo Chemical Co., Ltd., Taiwan), and rested for 10 min. The substrate was then heated at 100 °C for 10 min and cooled down to room temperature, thus completing the first screen-printing process. A second screen-printing process was then conducted through the repetition of the aforementioned steps on the same substrate. After the substrate was printed with the TiO_2 paste twice and cooled to room temperature, the substrate was placed on the screen printer and printed with a TiO_2 paste with a 200-nm particle diameter twice to fabricate a scattering layer. The working electrode was placed into a high-temperature furnace and thermal annealed at 500 °C for 30 min, cooled to room temperature, and immersed in the N719 dye solution for 24 h to adsorb the dye. The surface of the working electrode was then cleansed of excess dye with alcohol, thus completing the production of the working electrode (active area: 0.16 cm^2).

To prepare the counter electrode, two holes for electrolyte injection with 0.9-mm diameters were created in the FTO conductive glass substrate with a drill. The upper and lower surfaces of the substrate, excluding the 1.65-cm^2 area at the center, were then covered with 3M tape. An appropriate amount of the Pt nanoparticle paste was printed evenly onto the substrate with a scraper. After the 3M tape was removed, the substrate was thermal annealed at 450 °C for 30 min and cooled to room temperature, thus completing the production of the counter electrode. Finally, the DSSC devices were packaged. A 60-μm sealing film was cut into a 2 cm × 2 cm square, and a circular hole with a 0.9-cm diameter was cut in the center. The sealing film was placed between the working electrode and the counter electrode, and the two electrodes were adhered by pressing them at 3 kg/cm^2 and at 130 °C in a hot-press for 90 s and cooled to room temperature. Subsequently, 7 μL of electrolyte solution (0.6 M 1-buty-3-methylimidazolium iodide, 0.05 M LiI, 0.03 M I_2, 0.5 M 4-tert-butylpyridine, 0.1 M guanidine thiocyanate, and 5:1 of an ACN–valeronitrile solution) was injected into the electrolyte injection holes. Finally, the sealing film and glasses were used to cover the injection holes, followed by pressing at 130 °C and at 3 kg/cm^2 for 15 s to prevent electrolyte leakage, thus completing the fabrication of the DSSC devices.

2.3. Characteristics of the Working Electrode and the DSSC Devices

In this study, the characteristics of the working electrode, DSSC devices, and electron transport were analyzed. A scanning electron microscope (SEM) was employed to examine the surface of the working electrode. An energy dispersive spectrometer (EDS) was incorporated to observe the elemental compositions and proportions of the electrode. An X-ray diffractometer (XRD) was applied to analyze the lattice structure of the electrode. An X-ray photoelectron spectrometer (XPS) was used to determine the elemental compositions and chemical bonds on the surface of the electrode. The characteristics of the DSSC devices were investigated using current density–voltage (J–V), incident photon-to-electron conversion efficiency (IPCE), and electrochemical impedance spectroscopy (EIS). A simulated AM 1.5 G solar light emitted by a 550-W Xenon lamp solar simulator was used to measure the current density–voltage (J–V) characteristics of the DSSCs. The intensity of the incident light was calibrated to 100 mW/cm^2 by employing a reference cell. The reference cell had been given certification Bunkoh-Keiki Co Ltd, Tokyo, Japan. An external bias voltage was applied to the cell, and the photocurrent resulting was measured to obtain the photocurrent–voltage curves. The open-circuit voltage (V_{OC}), short-circuit current density (J_{SC}), fill factor (FF), and power conversion efficiency (PCE) for the DSSCs were extracted from the J–V characteristics. The electron transport within the DSSC was examined using intensity-modulated photocurrent spectroscopy (IMPS) and intensity-modulated photovoltage spectroscopy (IMVS).

3. Results and Discussion

3.1. Characteristics of the Working Electrode

Figure 1a shows the surface morphology of the TiO_2 working electrode as observed through the SEM. The TiO_2 nanoparticles formed an even, thin film on the FTO conductive glass. The mesoporous structure of the thin film exhibited a high surface area and enhanced the N719 dye adsorption and photocurrent generation, thereby reinforcing the photoelectron conversion efficiency. The diameters of the TiO_2 nanoparticles were 15–20 nm. Figure 1b–d shows the surfaces of the TiO_2–NiO (99:1), TiO_2–NiO (98:2), and TiO_2–NiO (97:3) working electrodes, respectively. All the working electrodes exhibited the same mesoporous TiO_2 structure regardless of whether or not they were mixed with NiO. However, some parts of the electrodes differed from others in their compositions and were relatively aggregated; this may have been caused by the NiO mixture and the surface characteristics of TiO_2 and NiO nanoparticles. Overall, NiO nanoparticles were evenly distributed on the TiO_2–NiO working electrodes. Previous studies have indicated that the thickness of a working electrode affects its photoelectron conversion efficiency, and an approximate thickness of 13 μm yields the optimal conversion efficiency. Therefore, this study investigated the effects of the NiO nanoparticles on the photoelectron conversion efficiency of the electrodes at the optimal thickness. Figure 2 shows the cross-sectional image of the working electrode as observed through the SEM. The lower layer was the TiO_2 active layer with a 7.125-μm thickness, and the upper layer was the large-particle TiO_2 scattering layer with a 6.563-μm thickness. The total thickness was 13.688 μm, satisfying the expectations of a working electrode. This study also investigated the adsorbed dye amounts on the working electrodes. The adsorbed N719 dye amounts on the TiO_2, TiO_2–NiO (99:1), TiO_2–NiO (98:2), and TiO_2–NiO (97:3) working electrodes were 126.2, 124.8, 123.6, and 122.3 nmol/cm^2, respectively. The adsorbed dye amounts decreased slightly following an increase in the proportion of NiO.

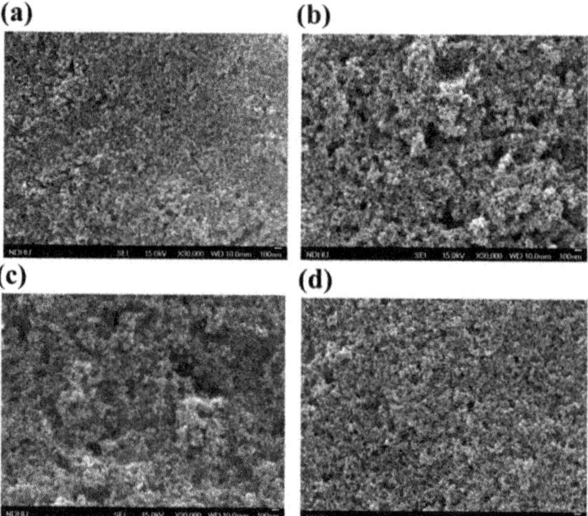

Figure 1. The SEM images of the (**a**) TiO_2, (**b**) TiO_2–NiO (99:1), (**c**) TiO_2–NiO (98:2), and (**d**) TiO_2–NiO (97:3) working electrodes.

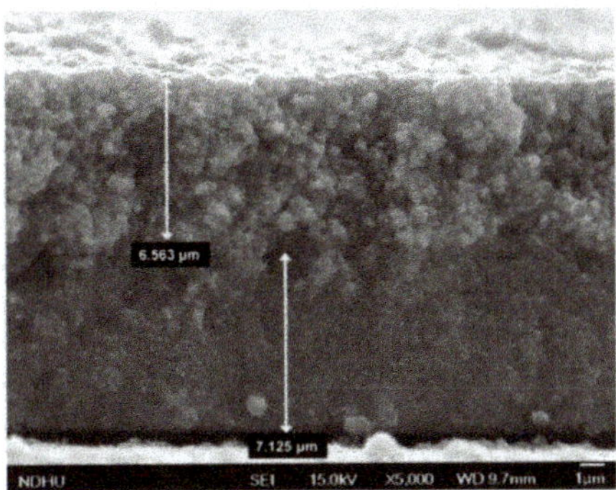

Figure 2. The SEM cross-sectional image of the working electrode.

Figure 3 shows the EDS analysis of the elemental compositions and proportions of the working electrode. As shown in Figure 3a, Ni did not exist on the electrode. Figure 3b–d shows the EDS analysis results of the TiO_2–NiO (99:1), TiO_2–NiO (98:2), and TiO_2–NiO (97:3) working electrodes, respectively. A rise in the proportion of NiO increased the amount of Ni on the electrode in both its weight (from 0.05% to 0.90%) and atomic ratios (from 0.02% to 0.40%), confirming the presence of NiO in the electrodes. The EDS results showed that the Ti to Ni ratio is different from the predicted proportions (i.e., 99:1, 98:2, and 97:3). This is because EDS cannot be considered a quantitative method of analysis; it can only be used as a qualitative analysis for various elements in an electrode. Figure 4 shows the XRD analysis results of the lattice structures of the TiO_2, TiO_2–NiO (99:1), TiO_2–NiO (98:2), and TiO_2–NiO (97:3) working electrodes. The characteristic peaks of the FTO were at 26.5°, 33.7°, 37.8°, 51.6°, 61.7°, and 65.6°. The characteristic peaks of the anatase TiO_2 were at 25.3° and 48.2°, and their respective corresponding lattice planes were (101) and (200). The characteristic peak of the rutile TiO_2 was at 54.7°, and its corresponding lattice plane was (211). The characteristic peak of NiO was at 43.3°, and its corresponding lattice plane was (200) [38–41]; an increase in the proportion of NiO raised the value of the characteristic peak. According to the XRD analysis, the diffraction peak of NiO was at 43.3°, and the NiO lattice structures were observed in the TiO_2–NiO (99:1), TiO_2–NiO (98:2), and TiO_2–NiO (97:3) working electrodes.

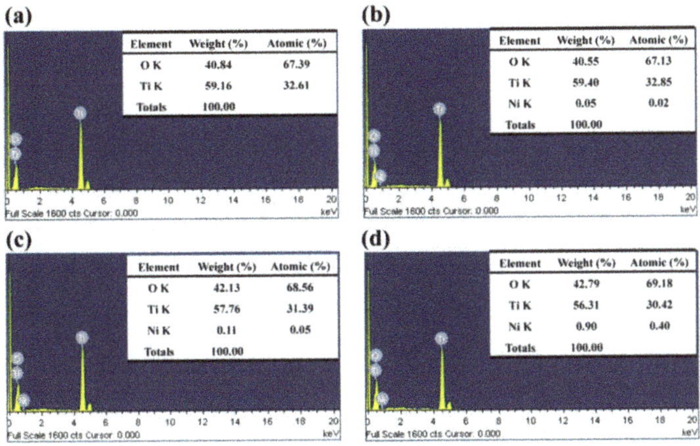

Figure 3. The EDS results of the (**a**) TiO_2, (**b**) TiO_2–NiO (99:1), (**c**) TiO_2–NiO (98:2), and (**d**) TiO_2–NiO (97:3) working electrodes.

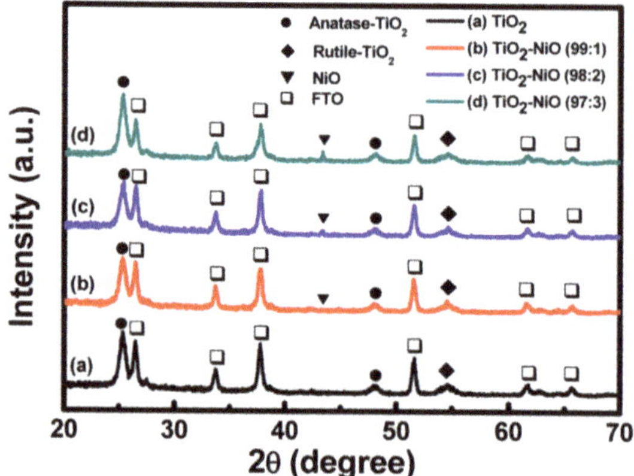

Figure 4. The XRD analysis results of various working electrodes.

Regarding the XPS analysis results of the elemental compositions and chemical bonds on the surfaces of the working electrodes, Figure 5a shows the O1s XPS analysis results of TiO_2 in the TiO_2–NiO (99:1), TiO_2–NiO (98:2), and TiO_2–NiO (97:3) electrodes. Figure 5b shows the Ti2p XPS analysis results. According to Figure 5a, no significant difference was observed among the electrodes regarding their O1s binding energy characteristic peak. However, the peak intensities of the electrodes with NiO were lower than those of the TiO_2 electrodes without NiO. The binding energy characteristic peak was at 529.8 eV, indicating the formation of an O^{2-} bond [42]. According to the Ti2p XPS analysis (Figure 5b), no significant difference was observed among the electrodes regarding their Ti2p binding-energy characteristic peaks; however, the peak intensity decreased following an increase in the proportion of NiO. This was because adding NiO led to a decrease in the proportion of TiO_2 and, subsequently, the Ti bond. The Ti bond energy could be divided into Ti $2p_{1/2}$ and Ti $2p_{3/2}$, which exhibited the characteristic peaks at 464.2 and 458.5 eV and which were both Ti^{4+} bonds [43]. Accordingly, the strength of the Ti binding-energy peak signals decreased following an increase in the proportion of NiO. Figure 6 displays the full spectrum XPS results of the TiO_2, TiO_2–NiO (99:1), TiO_2–NiO (98:2), and TiO_2–NiO (97:3) working electrodes. The signals at 285, 458.5, 529.8, and 566.4 eV correspond to the C1s, Ti2p, O1s, and Ti2s peaks, respectively. Because the amount of NiO added in the TiO_2–NiO electrode was nonsignificant, no Ni bonds were detected on the surfaces of the electrodes.

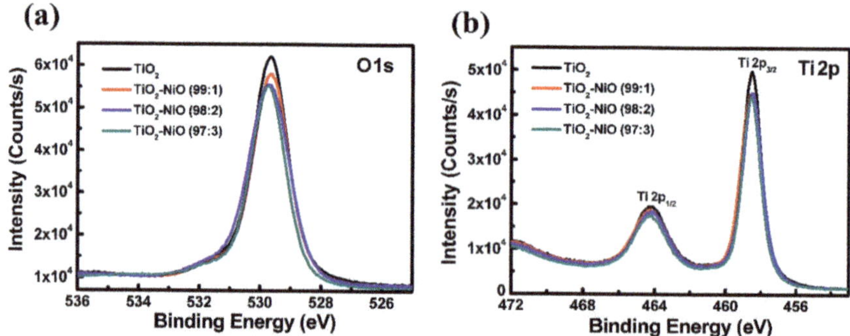

Figure 5. The (**a**) O1s and (**b**) Ti2p XPS analysis results of various working electrodes.

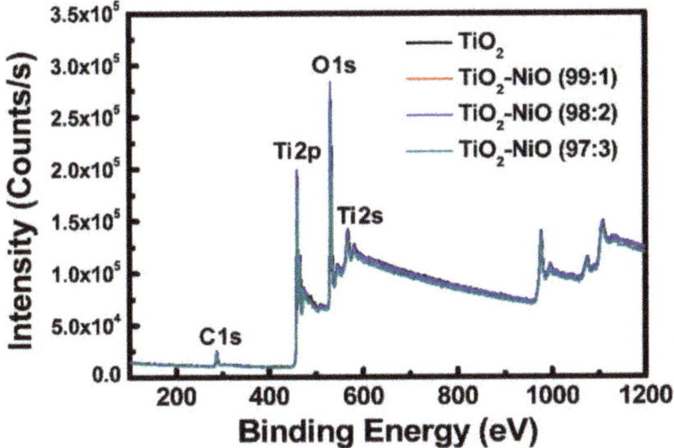

Figure 6. The full spectrum XPS analysis results of various working electrodes.

3.2. Characteristics of the DSSC Devices

In this study, the $J–V$ curves, IPCE, and EIS of the DSSC devices were further analyzed. Figure 7 shows the $J–V$ curves of the DSSCs. The DSSC's photoelectric characteristics, including the short-circuit current density (J_{SC}), open circuit voltage (V_{OC}), fill factor (FF), and power conversion efficiency (PCE) are listed in Table 1. The reproducibility of the device performances based on various working electrodes was investigated. For each condition, four DSSC devices were fabricated and analyzed. The values shown in Table 1 are averaged values and standard deviations of four devices made under identical conditions.

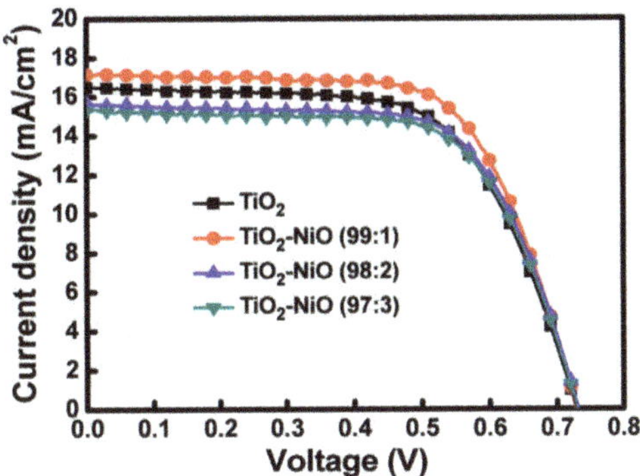

Figure 7. The $J–V$ curves of the dye-sensitized solar cells (DSSCs) based on various working electrodes.

Table 1. Characteristics of DSSCs fabricated using various working electrodes.

Working Electrode	J_{SC} (mA/cm^2)	V_{OC} (V)	Fill Factor	PCE (%)
TiO$_2$	16.49 ± 0.05	0.73 ± 0.01	0.64	7.70 ± 0.13
TiO$_2$–NiO (99:1)	17.15 ± 0.05	0.73 ± 0.01	0.67	8.39 ± 0.14
TiO$_2$–NiO (98:2)	15.69 ± 0.04	0.73 ± 0.01	0.67	7.67 ± 0.13
TiO$_2$–NiO (97:3)	15.34 ± 0.05	0.73 ± 0.01	0.66	7.39 ± 0.12

According to the experimental results, the J_{SC} of the TiO$_2$ working electrode was 16.49 mA/cm^2 without NiO, which was increased to 17.15 mA/cm^2 when NiO was added to TiO$_2$–NiO (99:1) and dropped to 15.69 mA/cm^2 and 15.34 mA/cm^2 at the respective ratios of 98:2 and 97:3. However, the V_{OC} was consistently 0.73 V throughout. The FFs of the working electrode were 0.64, 0.67, 0.67, and 0.66 when the TiO$_2$–NiO ratios were 100:0, 99:1, 98:2, and 97:3, respectively; an increase in the proportion of

NiO led to a rise in the FF until the TiO$_2$–NiO ratio was 97:3, at which point the FF started to decrease. The PCEs were 7.70%, 8.39%, 7.67%, and 7.39% at the TiO$_2$–NiO ratios of 100:0, 99:1, 98:2, and 97:3, respectively. According to the J–V curve analysis, adding an appropriate amount of NiO increased the PCE of the DSSC by 8.96%. In a DSSC, the electrons in the conduction band of TiO$_2$ might recombine with the dye and electrolyte. The p-type semiconductor material of NiO exhibits an energy level (−2.36 eV vs. the normal hydrogen electrode) higher than that of TiO$_2$ (−0.25 eV vs. the normal hydrogen electrode) [44]. The energy-level difference between NiO and TiO$_2$ formed a potential barrier that prevented the recombination of the electrons in the TiO$_2$ conduction band with the I$_3^-$ ions in the electrolyte. This also enabled the electrons to move to the external circuit, thus improving the PCE and J$_{SC}$. However, when an excessive amount of NiO was added, potential barriers prevented the electrons from moving to the external circuit, thus inhibiting the PCE and J$_{SC}$.

According to the IPCE analysis, the range of wavelengths was 300–800 nm (Figure 8). The IPCE peak value (88.3%) of the TiO$_2$–NiO (99:1) working electrode was identified at 540 nm, the highest IPCE value, followed by the peak values of pure TiO$_2$, TiO$_2$–NiO (98:2), and TiO$_2$–NiO (97:3). The IPCE value was positively correlated to J$_{SC}$. Therefore, the overall IPCE analysis result was consistent with the changes in the J$_{SC}$ in the J–V curve analysis. The IPCE value was highest at the wavelength range of 400–650 nm, which was the range for N719 dye absorption. At 340 nm, each of the devices mixed with NiO exhibited an absorption peak and those without NiO did not. This was because the 3.6 eV [45] bandgap of NiO corresponded to the absorption wavelength of 344 nm in the photon energy equation, which was consistent with the IPCE analysis results.

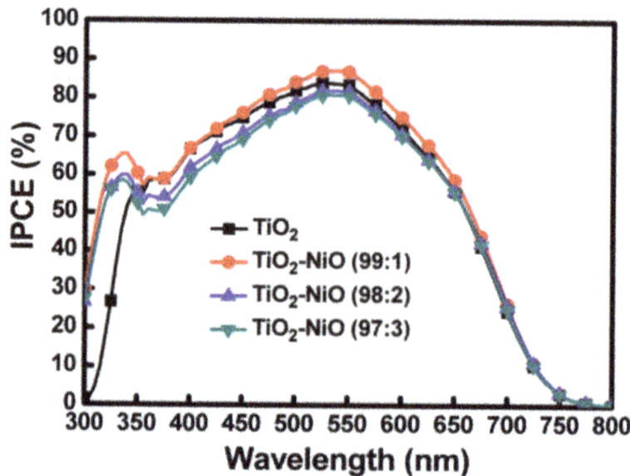

Figure 8. The incident photon-to-electron conversion efficiency (IPCE) results of the DSSCs based on various working electrodes.

The EIS analysis results of the DSSCs are shown in Figure 9. The range of measurement frequencies was 0.1 Hz to 100 kHz, and the AC amplitude was 0.01 V. EIS assesses the charge transfer processes in DSSCs, such as electron transport at the Pt/electrolyte interface (R_{CE}), electron transport and charge recombination at the TiO$_2$/dye/electrolyte interface (R_{ct}), I$_3^-$ transport in the electrolyte (W_D), and series resistance associated with the contribution from the FTO and counter electrodes (R_s). To extract the quantitative impedance characteristics of the devices, an equivalent circuit model (in Figure 9) was used to analyze the internal impedance of the devices. The extracted quantitative impedance parameters are listed in Table 2. The EIS results showed that the charge transfer resistance at the TiO$_2$/dye/electrolyte interface (Rct) was minimal when no NiO was added and increased marginally following an increase in the proportion of NiO. Accordingly, NiO inhibited the electron transport more effectively than TiO$_2$, and adding NiO in the TiO$_2$ working electrode slightly increased the resistance of the electron transport. Furthermore, the energy-level difference between NiO and TiO$_2$ formed a potential barrier that prevented the recombination of the electrons in the TiO$_2$ conduction band with the I$_3^-$ ions in the electrolyte, thereby increasing the charge transfer impedance at the TiO$_2$/dye/electrolyte interface.

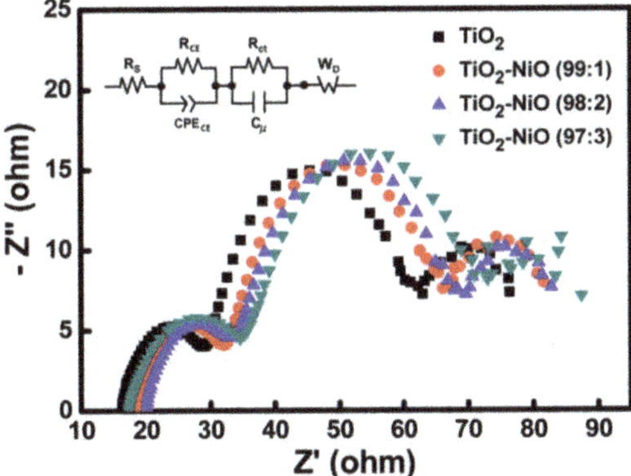

Figure 9. The electrochemical impedance spectroscopy (EIS) results of the DSSCs based on various working electrodes.

Table 2. The extracted quantitative impedance parameters from the EIS Nyquist plots of the DSSCs based on various working electrodes.

Working Electrode	R_S (ohm)	R_{CE} (ohm)	R_{ct} (ohm)	W_D (ohm)
TiO$_2$	16.12	10.74	29.94	13.28
TiO$_2$–NiO (99:1)	18.32	11.04	30.44	13.73
TiO$_2$–NiO (98:2)	19.61	11.26	31.48	14.14
TiO$_2$–NiO (97:3)	17.15	11.78	32.50	14.39

3.3. Characteristics of the Electron Transport

Figure 10 and Table 3 show the results of the IMPS and IMVS analyses, which were conducted to further clarify the characteristics of the electron transport in the DSSC. The measured data were presented on a complex plane, and the frequency corresponding to the imaginary minimal value was applied to calculate the electron transit time (τ_d) and the electron lifetime (τ_n) by the formulas $\tau_d = 1/2\pi f_{min}$ (IMPS) and $\tau_n = 1/2\pi f_{min}$ (IMVS). After obtaining the electron transit time (τ_d) and lifetime (τ_n), the diffusion coefficient (D) can be calculated by the formula $D = L^2/\tau_d$, where L is the film thickness of the working electrode. The charge collection efficiency (η_{CC}) can be calculated by the formula $\eta_{CC} = 1-\tau_d/\tau_n$ [46]. According to the IMPS analysis, the electron transit times of the working electrode were 14.67, 15.84, 23.25, and 54.09 ms at the TiO$_2$–NiO ratios of 100:0, 99:1, 98:2, and 97:3, respectively. This revealed that adding NiO to the electrode caused a decrease in the amount of TiO$_2$ (which facilitated electron transport) and an increase in the transport distance, thus extending the electron transit time. The mechanism of electron transport in TiO$_2$ and TiO$_2$–NiO working electrodes is shown in Figure 11. According to the IMVS analysis, the electron lifetimes of the working electrode were 116.53, 135.86, 100.42, and 92.56 ms at the TiO$_2$–NiO ratios of 100:0, 99:1, 98:2, and 97:3, respectively. At the 99:1 ratio, a small amount of NiO generated a potential barrier through its energy-level difference with TiO$_2$ and thereby reduced the electron recombination, thus maximizing the electron lifetime in the process. An excessive amount of NiO caused an increase in the electron transport resistance and transport distance, thereby leading to a reduction in the electron lifetime.

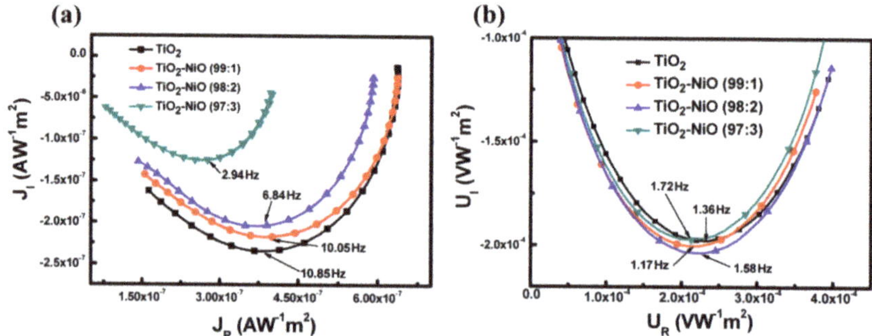

Figure 10. The (**a**) intensity-modulated photocurrent spectroscopy (IMPS) and (**b**) intensity-modulated photovoltage spectroscopy (IMVS) results of the DSSCs based on various working electrodes.

Table 3. Characteristics of the electron transit time (τ_d), lifetime (τ_n), diffusion coefficient (D), and collection efficiency (η_{cc}) of DSSCs with various working electrodes measured by IMPS and IMVS.

Working Electrode	τ_d (ms)	τ_n (ms)	D (m^2/s)	η_{cc}
TiO$_2$	14.67	116.53	3.19×10^{-9}	0.87
TiO$_2$–NiO (99:1)	15.84	135.86	2.96×10^{-9}	0.88
TiO$_2$–NiO (98:2)	23.25	100.42	2.01×10^{-9}	0.77
TiO$_2$–NiO (97:3)	54.09	92.56	8.66×10^{-10}	0.42

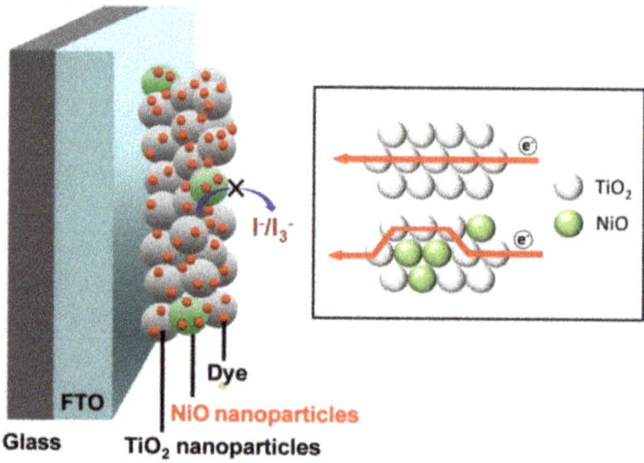

Figure 11. The mechanism of electron transport in TiO$_2$ and TiO$_2$–NiO working electrodes.

The thickness of the film on the working electrode (13.688 μm), measured according to the cross-sectional SEM image, was applied to calculate the electron diffusion coefficients, which were 3.19×10^{-9}, 2.96×10^{-9}, 2.01×10^{-9}, and 8.66×10^{-10} m^2/s when the TiO$_2$–NiO ratios were 100:0, 99:1, 98:2, and 97:3, respectively, confirming that NiO inhibited electron transport. The charge collection efficiencies (η_{CC}) were 0.87, 0.88, 0.77, and 0.42 when the TiO$_2$–NiO ratios were 100:0, 99:1, 98:2, and 97:3, respectively. Adding NiO in the electrode inhibited the electron transport and reduced the electron recombination. When the TiO$_2$–NiO ratio was 99:1, recombination was reduced without affecting the electron transport excessively, thereby optimizing the η_{CC}. The η_{CC} is an important factor affecting the Jsc of the DSSCs [47]. The Jsc values of the DSSCs were consistent with the trend of the η_{CC} values. Furthermore, the FF values of the DSSCs were not apparently affected by the η_{CC} values. This is because the FF is mainly determined by the internal resistance of the devices [48], and the internal resistance of the DSSCs only slightly increased after adding NiO in the TiO$_2$ working electrode.

4. Conclusions

In this study, the effects of the proportion of NiO in the TiO_2 solution on the TiO_2 working electrode, DSSC devices, and electron transport were investigated. In the SEM analysis of the surface of the working electrode, the TiO_2–NiO working electrodes exhibited a more aggregated morphology than that of the TiO_2 working electrodes. The thickness of the electrode was confirmed by the cross-sectional SEM analysis. According to the EDS analysis results, the weight and atomic ratios of Ni in the electrode increased following a rise in the amount of NiO added. The XRD analysis revealed that the TiO_2 in the working electrode was anatase TiO_2, and the diffraction peak of NiO was observed in the TiO_2–NiO working electrodes. According to the XPS analysis results, the binding-energy characteristic peaks of the O and Ti bonds did not change substantially following the increase in the amount of NiO added; however, the peak intensities decreased after the addition of NiO. Regarding the characteristics of the DSSC devices, according to the J–V curve analysis, the J_{SC} and PCE were optimal when the ratio of TiO_2–NiO was 99:1. Similarly, the IPCE peak value was optimized when the TiO_2–NiO ratio was 99:1. The IPCE peak formed by NiO was observed at a frequency of 340 nm. In the EIS analysis, the resistance corresponding to the electron transport at the TiO_2/dye/electrolyte interface increased with more NiO added, revealing that NiO inhibited electron transport. Moreover, the energy-level difference between TiO_2 and NiO generated a potential barrier that prevented the recombination of the electrons in the TiO_2 conduction band with the I_3^- ions in the electrolyte. According to the IMPS and IMVS analysis results, the electron transit time was extended following an increase in the amount of NiO added, confirming that NiO inhibited electron transport. The electron lifetime was optimized when the TiO_2–NiO ratio was 99:1; the potential barrier formed by TiO_2 and NiO through their energy-level difference prevented electron recombination, thereby prolonging the electron lifetime. The electron diffusion coefficient decreased following an increase in the amount of NiO added, further supporting the evidence that NiO inhibits electron transport. The charge collection efficiency was maximized when the TiO_2–NiO ratio was 99:1. Adding NiO to the TiO_2 working electrode both inhibited electron transport (a negative effect) and prevented electron recombination with the electrolyte (a positive effect). When the TiO_2–NiO ratio was 99:1, the positive effects outweighed the negative effects. Therefore, this was the optimal TiO_2–NiO ratio in the electrode for increasing the DSSC device efficiency and electron transport.

Author Contributions: Conceptualization, C.-H.T.; methodology, C.-H.T., C.-M.L. and Y.-C.L.; validation, C.-H.T.; formal analysis, C.-H.T., C.-M.L. and Y.-C.L.; investigation, C.-H.T., C.-M.L. and Y.-C.L.; resources, C.-H.T.; data curation, C.-H.T., C.-M.L. and Y.-C.L.; writing—original draft preparation, C.-H.T.; writing—review and editing, C.-H.T.; supervision, C.-H.T.; project administration, C.-H.T.; funding acquisition, C.-H.T. All authors have read and agreed to the published version of the manuscript.

Funding: The authors gratefully acknowledge the financial support from the Ministry of Science and Technology of Taiwan (MOST 108-2221-E-259-011).

Conflicts of Interest: The authors declare no conflict of interest.

References

1. O'Regan, B.; Grätzel, M. A low-cost, high-efficiency solar cell based on dye-sensitized colloidal TiO_2 films. *Nature* **1991**, *353*, 737–740.

2. Hagfeldt, A.; Grätzel, M. Light-induced redox reactions in nanocrystalline systems. *Chem. Rev.* **1995**, *95*, 49–68. [CrossRef]

3. Jiu, J.; Isoda, S.; Wang, F.; Adachi, M. Dye-sensitized solar cells based on a single-crystalline TiO_2 nanorod film. *J. Phys. Chem. B* **2006**, *110*, 2087–2092. [CrossRef] [PubMed]

4. Nazeeruddin, M.K.; Humphry-Baker, R.; Liska, P.; Grätzel, M. Investigation of sensitizer adsorption and the influence of protons on current and voltage of a dye-sensitized nanocrystalline TiO_2 solar cell. *J. Phys. Chem. B* **2003**, *107*, 8981–8987. [CrossRef]

5. Grätzel, M. Photoelectrochemical cells. *Nature* **2001**, *414*, 338–344. [CrossRef] [PubMed]

6. Hagfeldt, A.; Grätzel, M. Molecular photovoltaics. *Acc. Chem. Res.* **2000**, *33*, 269–277. [CrossRef]

7. Ting, H.C.; Tsai, C.H.; Chen, C.H.; Lin, L.Y.; Chou, S.H.; Wong, K.T.; Huang, T.W.; Wu, C.C. A novel amine-free dianchoring organic dye for efficient dye-sensitized solar cells. *Org. Lett.* **2012**, *14*, 6338–6341. [CrossRef]

8. Bessho, T.; Zakeeruddin, S.M.; Yeh, C.Y.; Diau, E.W.G.; Grätzel, M. Highly efficient mesoscopic dye-sensitized solar cells based on donor–acceptor-substituted porphyrins. *Angew. Chem. Int. Ed.* **2010**, *49*, 6646–6649. [CrossRef]

9. Chiba, Y.; Islam, A.; Watanabe, Y.; Komiya, R.; Koide, N.; Han, L. Dye-sensitized solar cells with conversion efficiency of 11.1%. *Jpn. J. Appl. Phys.* **2006**, *45*, L638. [CrossRef]

10. Nazeeruddin, M.K.; Zakeeruddin, S.M.; Humphry-Baker, R.; Jirousek, M.; Liska, P.; Vlachopoulos, N.; Shklover, V.; Fischer, C.-H.; Grätzel, M. Acid–base equilibria of (2,2′-Bipyridyl-4,4′-dicarboxylic acid)ruthenium(II) complexes and the effect of protonation on charge-transfer sensitization of nanocrystalline titania. *Inorg. Chem.* **1999**, *38*, 6298–6305. [CrossRef]

11. Chen, C.Y.; Wang, M.; Li, J.Y.; Pootrakulchote, N.; Alibabaei, L.; Ngoc-le, C.; Decoppet, J.; Tsai, J.; Grätzel, C.; Wu, C.G.; et al. Highly efficient light-harvesting ruthenium sensitizer for thin-film dye-sensitized solar cells. *ACS Nano* **2009**, *3*, 3103–3109. [CrossRef] [PubMed]

12. Chen, C.L.; Teng, H.; Lee, Y.L. In situ gelation of electrolytes for highly efficient gel-state dye-sensitized solar cells. *Adv. Mater.* **2011**, *23*, 4199–4204. [CrossRef] [PubMed]

13. Tsai, C.H.; Hsu, S.Y.; Lu, C.Y.; Tsai, Y.T.; Huang, T.W.; Chen, Y.F.; Jhang, Y.H.; Wu, C.C. Influences of textures in Pt counter electrode on characteristics of dye-sensitized solar cells. *Org. Electron.* **2012**, *13*, 199–205. [CrossRef]

14. Zhang, Q.; Dandeneau, C.S.; Zhou, X.; Cao, G. ZnO nanostructures for dye-sensitized solar cells. *Adv. Mater.* **2009**, *21*, 4087–4108. [CrossRef]

15. Duong, T.T.; Choi, H.J.; He, Q.J.; Le, A.T.; Yoon, S.G. Enhancing the efficiency of dye sensitized solar cells with an SnO_2 blocking layer grown by nanocluster deposition. *J. Alloys. Compd.* **2013**, *561*, 206–210. [CrossRef]

16. Congiu, M.; Marco, M.L.D.; Bonomo, M.; Nunes-Neto, O.; Dini, D.; Graeff, C.F.O. Pristine and Al-doped hematite printed films as photoanodes of p-type dye-sensitized solar cells. *J. Nanopart. Res.* **2017**, *19*, 7. [CrossRef]

17. Barea, E.; Xu, X.; González-Pedro, V.; Ripollés-Sanchis, T.; Fabregat-Santiago, F.; Bisquert, J. Origin of efficiency enhancement in Nb_2O_5 coated titanium dioxide nanorod based dye sensitized solar cells. *Energy Environ. Sci.* **2011**, *4*, 3414–3419. [CrossRef]

18. Memming, R.; Tributsch, H. Electrochemical investigations on the spectral sensitization of gallium phosphide electrodes. *J. Phys. Chem.* **1971**, *75*, 562–570. [CrossRef]

19. Wang, Z.S.; Kawauchi, H.; Kashima, T.; Arakawa, H. Significant influence of TiO_2 photoelectrode morphology on the energy conversion efficiency of N719 dye-sensitized solar cell. *Coord. Chem. Rev.* **2004**, *248*, 1381–1389. [CrossRef]

20. Yoon, J.H.; Jang, S.R.; Vittal, R.; Lee, J.; Kim, K.J. TiO_2 nanorods as additive to TiO_2 film for improvement in the performance of dye-sensitized solar cells. *J. Photochem. Photobiol. A* **2006**, *180*, 184–188. [CrossRef]

21. Liu, B.; Aydil, E.S. Growth of oriented single-crystalline rutile TiO_2 nanorods on transparent conducting substrates for dye-sensitized solar cells. *J. Am. Chem. Soc.* **2009**, *131*, 3985–3990. [CrossRef] [PubMed]

22. Feng, X.; Shankar, K.; Varghese, O.K.; Paulose, M.; Latempa, T.J.; Grimes, C.A. Vertically aligned single crystal TiO_2 nanowire arrays grown directly on transparent conducting oxide coated glass: Synthesis details and applications. *Nano Lett.* **2008**, *8*, 37813786.

23. Mor, G.K.; Shankar, K.; Paulose, M.; Varghese, O.K.; Grimes, C.A. Use of highly-ordered TiO_2 nanotube arrays in dye-sensitized solar cells. *Nano Lett.* **2006**, *6*, 215–218. [CrossRef] [PubMed]

24. Wei, Z.; Yao, Y.; Huang, T.; Yu, A. Solvothermal growth of well-aligned TiO_2 nanowire arrays for dye-sensitized solar cell: Dependence of morphology and vertical orientation upon substrate pretreatment. *Int. J. Electrochem. Sci.* **2011**, *6*, 1871–1879.

25. Li, Z.; Yu, L. The Size Effect of TiO_2 hollow microspheres on photovoltaic performance of ZnS/CdS quantum dots sensitized solar cell. *Materials* **2019**, *12*, 1583. [CrossRef]

26. Mahmoudabadi, Z.D.; Eslami, E. One-step synthesis of CuO/TiO_2 nanocomposite by atmospheric microplasma electrochemistry – Its application as photoanode in dye-sensitized solar cell. *J. Alloys Compd.* **2019**, *793*, 336–342. [CrossRef]

27. Chava, R.K.; Lee, W.M.; Oh, S.Y.; Jeong, K.U.; Yu, Y.T. Improvement in light harvesting and device performance of dye sensitized solar cells using electrophoretic deposited hollow TiO_2 NPs scattering layer. *Sol. Energ. Mat. Sol. C.* **2017**, *161*, 255–262. [CrossRef]

28. Haque, S.A.; Palomares, E.; Cho, B.M.; Green, A.N.M.; Hirata, N.; Klug, D.R.; Durrant, J.R. Charge separation versus recombination in dye-sensitized nanocrystalline solar cells: The minimization of kinetic redundancy. *J. Am. Chem. Soc.* **2004**, *127*, 3456–3462. [CrossRef]

29. Yu, H.; Zhang, S.; Zhao, H.; Will, G.; Liu, P. An efficient and low-cost TiO_2 compact layer for performance improvement of dye-sensitized solar cells. *Electrochim. Acta* **2009**, *54*, 1319–1324. [CrossRef]

30. Chou, C.S.; Yang, R.Y.; Yeh, C.K.; Lin, Y.J. Preparation of TiO_2/Nano-metal composite particles and their applications in dye-sensitized solar cells. *Powder Technol.* **2009**, *194*, 95–105. [CrossRef]

31. Habibi, M.H.; Karimi, B.; Zendehdel, M.; Habibi, M. Fabrication, characterization of two nano-composite CuO–ZnO working electrodes for dye-sensitized solar cell. *Spectrochim. Acta. A Mol. Biomol. Spectrosc.* **2013**, *116*, 374–380. [CrossRef] [PubMed]

32. Din, M.I.; Rani, A. Recent advances in the synthesis and stabilization of nickel and nickel oxide nanoparticles: A green adeptness. *Int. J. Anal. Chem.* **2016**, *2016*, 3512145.

33. Danial, A.S.; Saleh, M.M.; Salih, S.A.; Awad, M.I. On the synthesis of nickel oxide nanoparticles by sol–gel technique and its electrocatalytic oxidation of glucose. *J. Power Sources* **2015**, *293*, 101–108. [CrossRef]

34. El-Kemary, M.; Nagy, N.; El-Mehasseb, I. Nickel oxide nanoparticles: Synthesis and spectral studies of interactions with glucose. *Mater. Sci. Semicond. Process.* **2013**, *16*, 1747–1752. [CrossRef]

35. Bonomo, M. Synthesis and characterization of NiO nanostructures: A review. *J. Nanopart. Res.* **2018**, *20*, 222. [CrossRef]

36. Bonomo, M.; Mariani, P.; Mura, F.; Carlo, A.D.; Dini, D. Nanocomposites of nickel oxide and zirconia for the preparation of photocathodes with improved performance in p-type dye-sensitized solar cells. *J. Electrochem. Soc.* **2019**, *166*, D290–D300. [CrossRef]

37. Bandara, J.; Pradeep, U.W.; Bandara, R.G.S.J. The role of n–p junction electrodes in minimizing the charge recombination and enhancement of photocurrent and photovoltage in dye sensitized solar cells. *J. Photochem. Photobiol. A* **2005**, *170*, 273–278. [CrossRef]

38. Liu, B.; Sun, Y.; Wang, X.; Zhang, L.; Wang, D.; Fu, Z.; Lin, Y.; Xie, T. Branched hierarchical photoanode of anatase TiO_2 nanotubes on rutile TiO_2 nanorod arrays for efficient quantum dot-sensitized solar cells. *J. Mater. Chem. A* **2015**, *3*, 4445–4452. [CrossRef]

39. Chae, S.Y.; Hwang, Y.J.; Joo, O.S. Role of HA additive in quantum dot solar cell with $Co[(bpy)_3]^{2+/3+}$-based electrolyte. *RSC Adv.* **2014**, *4*, 26907–26911. [CrossRef]

40. Hsu, S.H.; Li, C.T.; Chien, H.T.; Salunkhe, R.R.; Suzuki, N.; Yamauchi, Y.; Ho, K.C.; Wu, K.C.W. Platinum-free counter electrode comprised of metal-organic-framework (MOF)-derived cobalt sulfide nanoparticles for efficient dye-sensitized solar cells (DSSCs). *Sci. Rep.* **2014**, *4*, 6983. [CrossRef]

41. Dharmaraj, N.; Prabu, P.; Nagarajan, S.; Kim, C.H.; Park, J.H.; Kim, H.Y. Synthesis of nickel oxide nanoparticles using nickel acetate and poly(vinyl acetate) precursor. *Mater. Sci. Eng. B* **2006**, *128*, 111–114. [CrossRef]

42. Mansour, A.N. Characterization of NiO by XPS. *Surf. Sci. Spectra* **1994**, *3*, 231. [CrossRef]

43. Diebold, U.; Madey, T.E. TiO_2 by XPS. *Surf. Sci. Spectra* **1996**, *4*, 227. [CrossRef]

44. Chou, C.S.; Lin, Y.J.; Yang, R.Y.; Liu, K.H. Preparation of TiO_2/NiO composite particles and their applications in dye-sensitized solar cells. *Adv. Powder Technol.* **2011**, *22*, 31–42. [CrossRef]

45. Kudo, A.; Miseki, Y. Heterogeneous photocatalyst materials for water splitting. *Chem. Soc. Rev.* **2009**, *38*, 253–278. [CrossRef]

46. Liao, J.Y.; Lei, B.X.; Kuang, D.B.; Su, C.Y. Tri-functional hierarchical TiO_2 spheres consisting of anatase nanorods and nanoparticles for high efficiency dye-sensitized solar cells. *Energy Environ. Sci.* **2011**, *4*, 4079–4085. [CrossRef]

47. Fakharuddin, A.; Ahmed, I.; Khalidin, Z.; Yusoff, M.M.; Jose, R. Channeling of electron transport to improve collection efficiency in mesoporous titanium dioxide dye sensitized solar cell stacks. *Appl. Phys. Lett.* **2014**, *104*, 053905. [CrossRef]

48. Liu, T.; Yu, K.; Gao, L.; Chen, H.; Wang, N.; Hao, L.; Li, T.; He, H.; Guo, Z. A graphene quantum dot decorated $SrRuO_3$ mesoporous film as an efficient counter electrode for high-performance dye-sensitized solar cells. *J. Mater. Chem. A* **2017**, *5*, 17848–17855. [CrossRef]

Article

Anti-Corrosive Properties of an Effective Guar Gum Grafted 2-Acrylamido-2-Methylpropanesulfonic Acid (GG-AMPS) Coating on Copper in a 3.5% NaCl Solution

Ambrish Singh [1,2,*], Mingxing Liu [1,2], Ekemini Ituen [1,2] and Yuanhua Lin [1,2,*]

[1] School of Materials Science and Engineering, Southwest Petroleum University, Chengdu 610500, China; m1763902986@163.com (M.L.); ebituen@gmail.com (E.I.)

[2] State Key Laboratory of Oil and Gas Reservoir Geology and Exploitation, Southwest Petroleum University, Chengdu 610500, China

* Correspondence: vishisingh4uall@gmail.com (A.S.); yhlin28@163.com (Y.L.); Tel.: +028-83037401 (Y.L.)

Received: 23 December 2019; Accepted: 24 February 2020; Published: 5 March 2020

Abstract: Guar gum grafted 2-acrylamido-2-methylpropanesulfonic acid (GG-AMPS) was synthesized using guar gum and AMPS as the base ingredients. The corrosion inhibition of copper was studied using weight loss, electrochemical, and surface characterization methods in a 3.5% sodium chloride (NaCl) solution. Studies including weight loss were done at different acid concentrations, different inhibitor concentrations, different temperatures, and different immersion times. The weight loss studies showed the good performance of GG-AMPS at a 600 mg/L concentration. This concentration was further used as the optimum concentration for all of the studies. The efficiency decreased with the rise in temperature and at higher concentrations of acidic media. However, the efficiency of the inhibition increased with the additional immersion time. Electrochemical methods including impedance and polarization were employed to calculate the inhibition efficiency. Both of the techniques exhibited a good inhibition by GG-APMS at a 600 mg/L concentration. Surface studies were conducted using scanning electrochemical microscopy (SECM), scanning electron microscopy (SEM), and atomic force microscopy (AFM) methods. The surface studies showed smooth surfaces in the presence of GG-AMPS and rough surfaces in its absence. The adsorption type of GG-AMPS on the surface of the copper followed the Langmuir adsorption model.

Keywords: corrosion; guar gum; coatings; electrochemical impedance spectroscopy (EIS); SECM; AFM

---

1. Introduction

Copper is used worldwide because of its good conductivity, availability, and large tenders. Almost all companies use copper in one way or another, because of its varied applications. Copper is an appropriate material for electrodes, wires, joints, and couplings [1]. An NaCl solution can cause severe corrosion in metals and alloys [2]. It can cause pitting corrosion under the coatings by forming blisters. These pits are difficult to detect under the coatings, and can cause fatal accidents and a shutdown of the systems. Known also as rock salt or common salt, sodium chloride is abundantly found in nature and mostly in sea water. Marine corrosion is very common and causes billions of dollars of losses globally. So, there is always a need to find solutions to marine corrosion.

The application of corrosion inhibitors is currently the most cost-efficient means of dealing with copper corrosion [3]. The corrosion inhibitors used at this stage are mainly organic corrosion inhibitors, because organic molecules may contain O, N, and S, as well as other atoms with unshared pair electrons and π-bonds [3]. Such organic molecules can be empty orbitals, which interact with copper surfaces [3]. The d-orbital acts to form a protective film to achieve corrosion protection. Glue extracted from natural plants is one of them. Gum contains a large amount of O atoms, and has received a lot of attention and has been widely studied. At the same time, the use of gums as corrosion inhibitors can effectively improve the utilization rate of natural plants. Chemical compounds or inhibitors are used to mitigate corrosion as one of the common available methods. These compounds are inorganic and organic in nature. They are rich in carbon, oxygen, nitrogen, sulfur, benzene rings, and double bonds in their molecular structures [4]. In the recent decade, the corrosion fraternity has been concerned with finding green and environment-friendly inhibitors [5]. Plant extracts, as potential corrosion inhibitors, have been used by several authors [6–10]. The shortcoming of using plant extracts is that they tend to develop fungi/bacteria on their surfaces after some time, and this affects their inhibition efficiency. So, the search for green polymers or biopolymers has been sought worldwide. Several polymers have been studied with heteroatoms in their structures [11–15]. The multifunctional groups in the polymers can easily adsorb or attach to any metal surface, thereby protecting them from corrosion. Polymers may prove to be better than the low molecular inhibitors used in the acidization process. Biocompatible compounds have also been used as inhibitors because of their cost effectiveness, ease of availability, and low-cost machines used. Our motivation was to develop a compound with all of the qualities of a polymer, in addition to being non-toxic and environmentally benign. Thus, guar gum grafted 2-acrylamido-2-methylpropanesulfonic acid (GG-AMPS) was synthesized as a green compound to cope with environmental regulations, and to be used effectively in high concentrations.

A survey of the literature reveals that no work has been done using GG-AMPS as a corrosion inhibitor in an NaCl solution. GG-AMPS has a nitrogen atom, oxygen atom, and sulfur atom in its molecular structure, which provide a good adsorption approach, leading to good bonding and complex grouping on the metal surface. The presence of hydroxyl groups at different points makes it a potential inhibitor that can share electrons and take part in good bond formation. This paper elucidates the inhibition effect of GG-AMPS for copper in a 3.5% NaCl media. The mitigation properties of GG-AMPS were conducted using static weight-loss methods and electrochemical methods. In the meantime, the surfaces of copper were examined by scanning electrochemical microscopy (SECM), scanning electron microscopy (SEM), and atomic force microscopy (AFM).

2. Experimental

2.1. Copper Samples

In all of the experiments, such as for the weight loss, the electrochemical and surface morphology of the pure copper samples were utilized. Each of the samples employed for weight loss were cut into rectangle coupons. Prior to the experiments, the surfaces were abraded with silicon sheets of grades 300 to 1200, cleaned with acetone and distilled water, and vacuum-dried. The dimensions of the copper samples employed for the weight loss tests were 2.0 cm × 2.5 cm × 0.2 cm.

2.2. Corrosive Medium

The inhibitor coatings were concentrated in the range of 100 to 600 mg/L for all of the studies. A 3.5% sodium chloride solution was used as the corrosive medium. It was prepared using pure NaCl and double-distilled water. A freshly prepared solution was used for each of the experiments. The different concentrations of inhibitor solutions were prepared by adding a calculated amount of inhibitor into the corrosive medium.

2.3. Synthesis of GG-AMPS

The synthesis of GG-AMPS was conducted according to the previous reference [16–18]. One gram of guar gum was slowly dissolved in 100 mL of distilled water. Then, 0.2 g of potassium persulfate was added to the guar gum solution, and the reaction continued for 1 h in a water bath at 70 °C. After that, 2 g of 2-acrylamide-2-methyl-1-propane sulfonic acid (AMPS) and 0.2 g of N, N′-methylenebisacrylamide were added to the above solution, and the reaction continued for 3 h at the same temperature (70 °C). The whole reaction process was carried out in a nitrogen atmosphere. After the solution was cooled, the excess acetone was added to the solution so as to separate the desired product. The precipitates were filtered and dried in vacuum at 50 °C for 24 h to a constant weight. The product was guar gum grafted with 2-acrylamide-2-methyl-1-propane sulfonic acid (GG-AMPS). The compound obtained was further characterized by infrared (IR) spectroscopy. The plan of the synthesis and molecular structure of the inhibitor coating is shown in Figure 1.

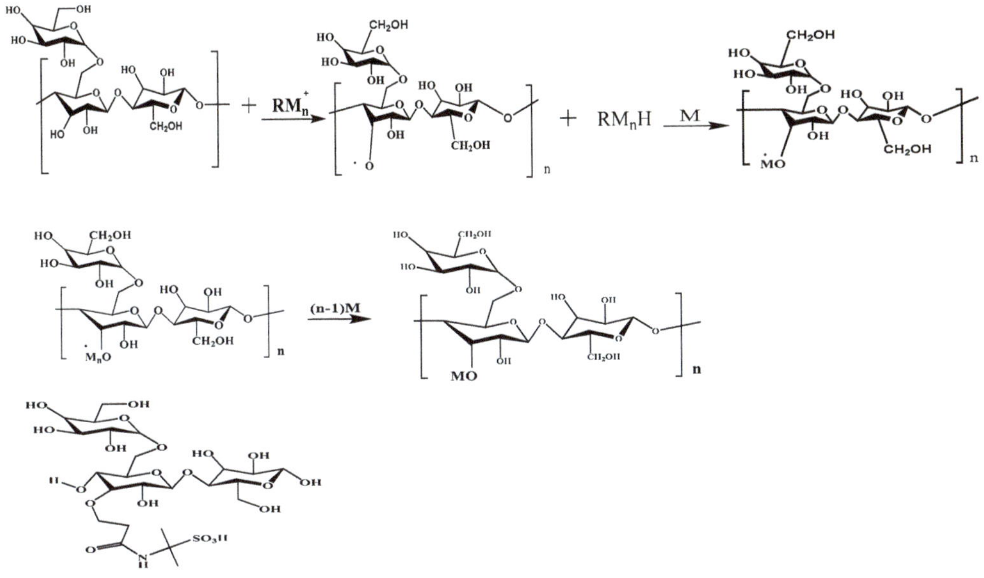

2-acrylamide-2-methyl-1-propane sulfonic acid (AMPS)

Figure 1. Molecular structure and synthesis scheme of guar gum grafted 2-acrylamido-2-methylpropanesulfonic (GG-AMPS). M = AMPS.

2.4. Infrared Spectroscopy

IR spectroscopy was conducted using the Nicolette 6700 infrared spectrometer of Thermo Electric Company Inc. from the USA (West Chester, PA). The infrared spectrum of the GG-AMPS is shown in Figure 2.

Figure 2 shows the IR spectrum of the guar gum and GG-AMPS. In Figure 2, 3461 cm^{-1} indicates the tensile vibration of O–H in the guar gum. In addition, the weak peak near 2926 cm^{-1} is the C–H, and the peak at 1638 cm^{-1} is the vibration peak of the six-membered rings. In Figure 2, the small peaks at 1552 and 1300 cm^{-1} are the bending vibrations of N–H in the AMPS amide group, and the wide peak at 3408 cm^{-1} is the overlap of the N–H stretching band and O–H stretching band, which resulted in a certain movement of the wide peak at 3461 cm^{-1} in the guar gum. The peak at 1657 cm^{-1} and its adjacent peak are the result of the overlapping of the C=O vibration and –CONH– vibration in –CO$_2$H. The peaks at 1375 and 1458 cm^{-1} are the C–H bending vibrations in –CH$_3$, –CH$_2$, and –CH. The peaks at 1220 and 1042 cm^{-1} are characteristic peaks of a sulfonic group. Figure 2 reveals that AMPS was successfully grafted with guar gum.

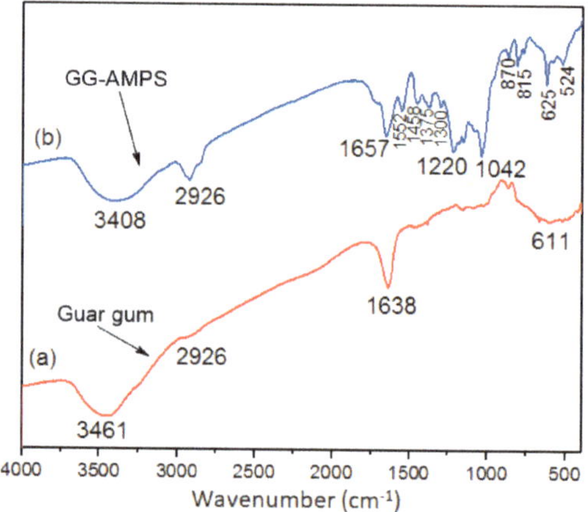

Figure 2. IR spectrum of GG-AMPS.

2.5. Weight Loss

The duration of all of the weight loss tests was determined following the ASTM G31-2004 standard [19]. The duration of the experiments selected for all of the tests was 24 h. The copper coupons were cleaned with water and then rinsed with Clarke's solution for 5 min. The coupons were then exposed to vacuum drying. The obtained weight loss values were used to calculate the corrosion rate of the metal in the corrosive media. Each of the tests were performed in triplicate and the mean values were reported. The corrosion rates were calculated using the following equation:

$$C_R(mm/y) = \frac{87.6W}{atD} \tag{1}$$

where W is the average weight loss of copper specimens (mg), a is total area of copper specimen, t is the immersion time (h), and D is the density of copper in (g·cm⁻³).

2.6. Electrochemical Analysis

A Gamry potentiostat workstation (Gamry, Warminster, PA, USA) was utilized for the electrochemical tests. The potentiostat was connected to a cell assembly, which consisted of a reference electrode, counter electrode, and working electrode. Prior to the start of the experiments, the working electrode (copper) was exposed to the 3.5% NaCl solution for 30 min, so as to keep the potential (E_{corr}) stable.

The range of the frequency selected was from 100 kHz to 10 mHz, at an amplitude of 10 mV per decade for all of the electrochemical impedance tests. The evaluation of the inhibition efficiency was done using the following equation:

$$\eta\% = \left(1 - \frac{R_{ct}}{R_{ct(i)}}\right) \times 100 \tag{2}$$

where R_{ct} and $R_{ct(i)}$ are the charge transfer resistances without and with GG-AMPS, respectively.

The potentiodynamic polarization tests were conducted in the limit of −250 to 250 mV, with a 0.167 mV/s scan rate. The following equation was used to determine the efficiency of the coating:

$$\eta\% = \left(1 - \frac{i_{corr(i)}}{i_{corr}}\right) \times 100 \tag{3}$$

where i_{corr} and $i_{corr(i)}$ are the corrosion current densities without and with GG-AMPS, respectively.

2.7. Scanning Electrochemical Microscopy (SECM)

The SECM tests were performed using a Princeton workstation equipped with Versa scan software (3000, Versa, TX, USA). The microprobe was made of a silver (Ag)/Pt wire inside a glass tube with a diameter of 10 μm. The probe vibrated over the metal surface at an average distance of 100 μm, and the scanned area was 20 μm × 20 μm.

2.8. Scanning Electron Microscopy (SEM)

The scanning electron microscopy (SEM) was done to detect the changes in the external area of the metal. The SEM was conducted using a Tescan machine (S800, Tescan, Shanghai, China) equipped with a Zeiss lens (Zeiss, Shanghai, China). The samples were washed with a sodium bicarbonate solution in order to remove the corrosion products, followed by distilled water prior to surface exposure.

2.9. Atomic Force Microscopy (AFM)

The AFM experiments were conducted using a Dimension Icon Brock instrument (HPI, Bruker, Karlsruhe), made in Germany, for all of the copper coupons. Once the tests were completed, the images obtained were sent to Nanoscope analysis software (2.0), version v1.40r1, so as to obtain the 3D figures. The average roughness and the peak roughness were further confirmed using the linear fitting of the 2D figures.

3. Results and Discussion

3.1. Weight Loss Experiment

3.1.1. Effect of Concentration, Time, and Temperature

The effect of the GG-AMPS concentrations on the protective covering of the copper surfaces is portrayed in the form of a concentration vs. inhibition efficiency graph (Figure 3a). From the figure, it is evident that the mitigation activity of GG-AMPS rose with the increase in concentration, and attained values of 95% at 600 mg/L. This increase in mitigation ability was due to the adsorption of the GG-AMPS molecules onto the wide area of the copper surface. Figure 3b displays the increase in inhibition efficiency with immersion time, for up to 12 h. This discovery points to the molecular structure of GG-AMPS having a big effect on the values of inhibition efficiency. Figure 3c depicts the influence of temperature on the inhibition efficiency of GG-AMPS. The efficiency was found to decrease with an increase in temperature. This may have been the result of the desorption of the coating from the copper surface. So, for this coating to be used at high temperatures, the concentration of the coating should be increased. Figure 3d displays the influence of the inhibition efficiency with the NaCl concentration. The efficiency was seen to decrease with the increase in NaCl concentration. This may be due to the NaCl solution penetrating the coating on the copper surface, causing pitting corrosion. In this study, the inhibitor molecules had π-electrons in the benzene ring and non-bonding electrons on the heteroatoms, such as oxygen, sulfur, and nitrogen, which assist the molecules in their adsorption onto the copper surface [20].

The adsorption of the coating on any metal surface also depends on the number of electron-donating functional groups attached to the structure. A greater number of electron-donating groups can help with better adsorption and bond formation, which can lead to a better mitigation of corrosion. GG-AMPS consists of OH, SO_3H, and NH groups, which may be a reason for its good mitigation abilities in the corrosion of copper in sodium chloride media.

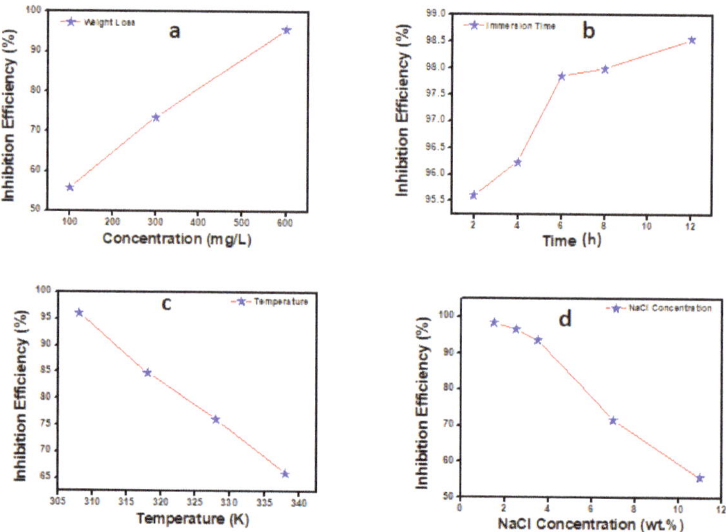

Figure 3. (**a**) Variation of inhibition efficiency (%) with inhibitor concentration. (**b**) Variation of inhibition efficiency (%) with immersion time. (**c**) Variation of inhibition efficiency (%) with temperature. (**d**) Variation of inhibition efficiency (%) with NaCl concentration.

3.1.2. Adsorption Isotherm of Inhibitor on Copper

The adsorption of molecules on the metal surface can be better explained using Langmuir, Frumkin, Flory Huggins, and Temkin isotherms. The obtained experimental data can be fit using the equations of these isotherms. The best fit gives a linear slope with regression coefficient values approaching unity. The fitted experimental values showed that Langmuir was the best out of all of the equations (Figure 4). The following equation was used to determine the fitted results of the isotherm [21]:

$$\frac{C_{\text{inh}}}{\theta} = \frac{1}{K_{\text{ads}}} + C_{\text{inh}} \tag{4}$$

where C_{inh} is the GG-AMPS concentration (mg/L), and θ and K_{ads} represent the surface coverage and adsorption constant, respectively.

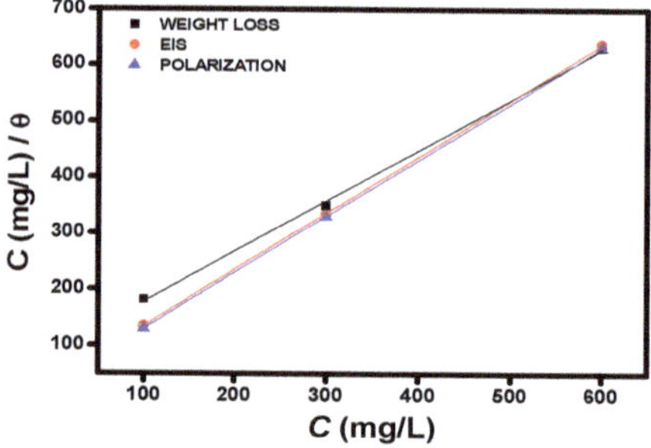

Figure 4. Langmuir adsorption isotherm plots using the values of weight loss, electrochemical impedance spectroscopy (EIS), and polarization.

3.2. Electrochemical Tests

3.2.1. Electrochemical Impedance Spectroscopy (EIS) Studies

The behavior of the copper electrode was investigated by electrochemical impedance tests. The impedance nature of the metal is represented in the form of Nyquist graphs (Figure 5a). As reported in Figure 5a, at a higher frequency, a capacitive loop is seen, which contains a straight line, tending to be a semicircle. The semicircle is normally the result of the capacitance of double-layer and charge-transfer resistance [22]. The presence of two time constants can be seen in the Nyquist and bode figures. The presence of two time constants may have been because of the roughness and inhomogeneity of the metal surface after corrosion. In addition, the diameter capacitance with GG-AMPS was bigger in comparison with the blank. Meanwhile, the diameter got bigger with higher concentrations of GG-AMPS. This was because the GG-AMPS molecules that were adsorbed on the copper surface formed a barrier and enhanced the corrosion resistance properties [23–25].

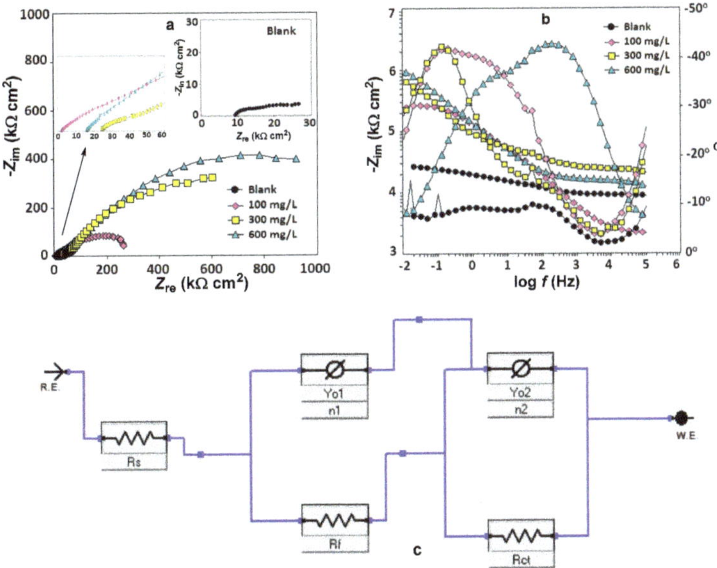

Figure 5. (**a**) Nyquist plots at different concentrations of GG-AMPS. (**b**) Bode, phase-angle plots at different concentrations of GG-AMPS. (**c**) Equivalent circuit used to fit and analyze the data.

In the bode plots (Figure 5b), the slope values tended to increase in the presence of GG-AMPS, rather than in its absence. This indicates the inhibition action of GG-AMPS on the copper surfaces. In the phase angle plots (Figure 5b), at the intermediate frequency, the pinnacle of the phase angle increased as the GG-AMPS concentration increased. The highest peak was observed at 43.2° for copper at a 600 mg/L inhibitor concentration. This was due to the corrosion mitigation on the copper surfaces by the GG-AMPS protective shield, which that isolated the copper from the corrosive media [26]. The binding of the GG-AMPS molecules with the metal surface was quite strong and stable, which finally enhanced its corrosion resistance quality. For the impedance data calculation, the circuit used is shown in Figure 5c. The circuit used was drawn using the model editor in Echem analyst. It contained charge-transfer resistance (R_{ct}), a solution resistor (R_s), film resistance (R_f), and two constant phase elements (CPEs). The CPEs were included in the circuit for the perfect fitting of the Nyquist curves, as they balance the deviation of surface roughness, disruption, imperfectness, impurity, and adsorption [27–30].

The CPE impedance is given below:

$$Z_{CPE} = Y_o^{-1}(iw)^{-n} \tag{5}$$

where Y_0, w, i, and n are the constant, angular frequency, an imaginary number, and an empirical exponent, respectively.

Table 1 shows the impedance values of the fitted curves. It can be seen that the R_{ct} and Y_0 parameters at all of the concentrations of GG-AMPS display a reverse pattern. This process is credited to the GG-AMPS molecules adsorbing onto the copper surface, which finally enhances the copper corrosion resistance attributes [31]. The value of R_{ct} at 600 mg/L for GG-AMPS is 905 kΩ·cm^2. Thus, GG-AMPS provides a good resistance to corrosive solutions. The rise in n values with the increase in GG-AMPS concentration was due to their adsorption and finally enhancement of their homogeneity [32]. Thus, with a decent number of heteroatom functional groups, the corrosion inhibition property increased.

Table 1. Electrochemical impedance parameters in the absence and presence of different concentrations of GG-AMPS at 308 K.

C_{inh}	R_s	R_{ct}	n_1	Y_{o1}	n_2	Y_{o2}	R_f	X^2	η_{EIS}
(mg/L)	(kΩ·cm^2)	(kΩ·cm^2)		(μF/cm^2)		(μF/cm^2)	(kΩ·cm^2)	$\times10^{-3}$	(%)
Blank	9.29	87	0.72	57.8	0.37	59.2	7.2	1.01	–
100	3.01	130	0.76	34.2	0.43	51.1	20.4	3.50	33.0
300	23.5	568	0.81	23.5	0.59	45.4	310.2	4.53	84.6
600	15.51	905	0.83	19.5	0.74	31.3	415.5	2.73	90.3

X^2 refers to Chi square.

3.2.2. Potentiodynamic Polarization Tests

The polarization plots of copper in NaCl solutions without and with different concentrations of GG-AMPS are represented in Figure 6. Several essential electrochemical factors, like the corrosion current density (i_{corr}), corrosion potential (E_{corr}), cathodic Tafel slope (β_c), anodic Tafel slope (β_c), and inhibition efficiency (η%), are tabulated in Table 2. The analysis of Table 2 suggests that the corrosion current density shifted from 98.7 μA/cm^2 (3.5% NaCl) to 4.9 μA/cm^2 (600 mg/L GG-AMPS), and this represents that the GG-AMPS coating was effective for the mitigation of corrosion. The value of maximal efficiency as obtained was 95% at 600 mg/L. As can be seen, after the inclusion of GG-AMPS in the corrosive media, both the anodic and cathodic current density were reduced. The addition of GG-AMPS in higher concentrations caused more H$^+$ ion reduction in the system than the anodic dissolution process. As is evident in Figure 6, the collateral cathodic slopes suggest the conversion of H$^+$ to H$_2$ was not varied. Similarly, with higher concentrations of GG-AMPS, the β_c values were shifted, suggesting that GG-AMPS affected the kinetics of H$_2$ evolution. This may be accredited to the diffusion or the shield phenomenon [33]. Likewise, the values of β_a also underwent a modification with the increase in the GG-AMPS concentration, suggesting that the designed coating primarily underwent adsorption all over the copper surface. This phenomenon shows the mitigation of the corrosion reaction by obstructing the activated centers, without modifying the mechanism of the anodic process [34].

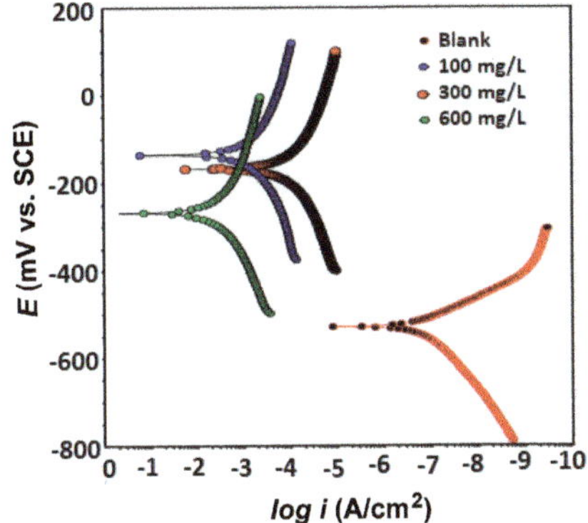

Figure 6. Potentiodynamic polarization curves in the absence and presence of different concentrations of GG-AMPS.

Table 2. Electrochemical polarization parameters in the absence and presence of different concentrations of GG-AMPS at 308 K.

Inhibitor	E_{corr}	i_{corr}	β_a	$-\beta_c$	η
(mg/L)	(mV/SCE)	(μA/cm^2)	(mV/dec)	(mV/dec)	(%)
Blank	−579	98.7	67	358	−
100	−145	21.2	267	79	78.5
300	−189	12.0	233	73	87.8
600	−293	4.9	199	84	95.0

In addition, the polarization curves showed that the addition of GG-AMPS mitigated both the cathodic and anodic processes. Therefore, the coating can be categorized into mixed forms. Nevertheless, the variations in the E_{corr} values in the presence of the coating were towards the anodic route, as compared with those without the coating, indicating that GG-AMPS is predominantly cathodic.

3.3. Scanning Electrochemical Microscopy (SECM)

Scanning electrochemical microscopy is very useful for detecting localized corrosion on the metal surface. Figure 7a,b shows the 2D and 3D pictures of the *x*-axis for copper without GG-AMPS in a seawater solution. A very high current was observed in the 2D maps and 3D structures (Figure 7a,b). This can be endorsed by the straight connection of the probe with the copper surface. As the probe was moved at a certain visible corroded part on the metal surface observed through the camera, the corrosion profile was detected. However, for the copper surface with the GG-AMPS film on it, the 2D and 3D maps showed a lower current (Figure 7c,d). A lower current was observed in the presence of the GG-AMPS film, which may have been due to the GG-AMPS coating on the copper surface forming a hydrophobic film that repelled the corrosive solution. This phenomenon indicated that the GG-AMPS film protected the copper surface well from corrosion in the NaCl solution.

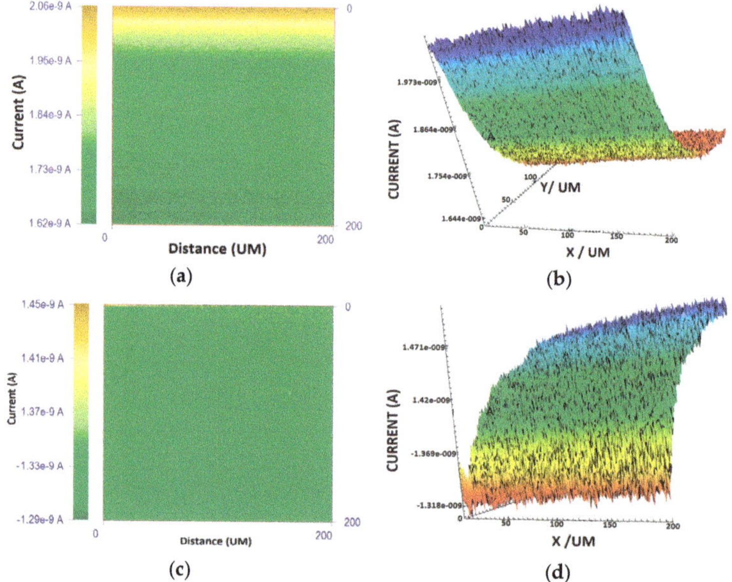

Figure 7. Scanning Electrochemical Microscopy (SECM) images of (**a**) 2D and (**b**) 3D copper surface in 3.5% NaCl solution and (**c**) 2D and (**d**) 3D copper surface coated with GG-AMPS in 3.5% NaCl solution.

3.4. Scanning Electron Microscopy (SEM)

The copper coupons with 600 mg/L GG-AMPS and 3.5% NaCl were exposed to SEM, as depicted in Figure 8. The surface morphology of the copper without GG-AMPS was very rough with several corrosion products, because of the rampant dissolution and deterioration of the metal (Figure 8a). Nevertheless, the addition of the GG-AMPS showed a smooth copper surface (Figure 8a). However, the surface showed a porous coating with abraded lines visible through them. This phenomenon suggested that, with GG-AMPS, the rate of corrosion was decreased due to the GG-AMPS coating having formed a conserving film over the copper surface.

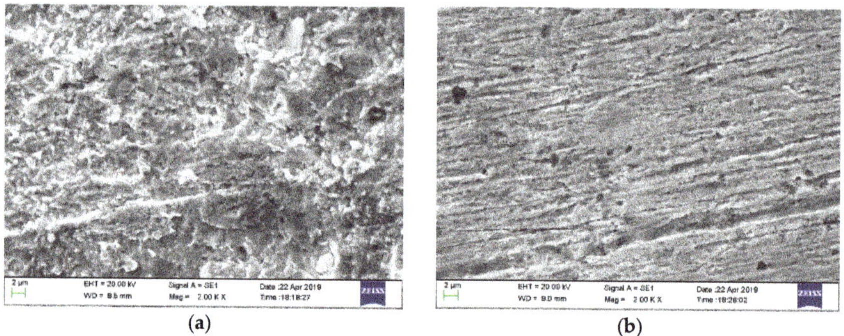

Figure 8. SEM images of a (**a**) copper surface in a 3.5% NaCl solution, and a (**b**) copper surface coated with GG-AMPS in a 3.5% NaCl solution.

3.5. Atomic Force Microscope (AFM)

The 3D micro-structural pictures of the copper surface are displayed in Figure 9. The clear indication of the terrible deterioration of the copper surface without coating due to corrosion can be seen in Figure 9a,b. Nevertheless, the roughness of the copper surface was comparatively decreased with the addition of GG-AMPS (Figure 9c,d). This anti-corrosive nature of the coating is surely because of the good binding with the metal surface.

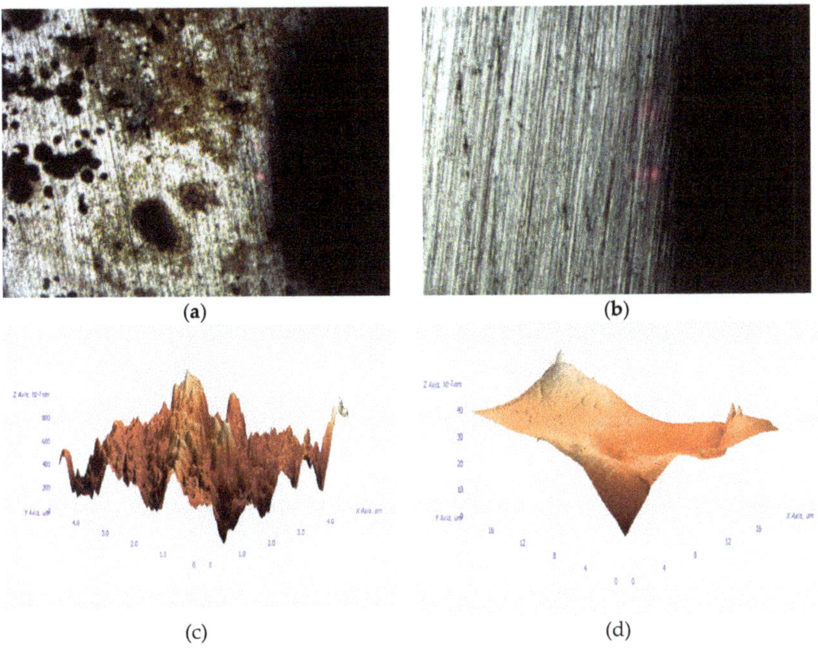

Figure 9. Optical image (**a**) copper surface in a 3.5% NaCl solution (**b**) copper surface coated with GG-AMPS in a 3.5% NaCl solution; AFM images (**c**) copper surface in a 3.5% NaCl solution (**d**) copper surface coated with GG-AMPS in a 3.5% NaCl solution.

4. Conclusions

The tested GG-AMPS coating is good for copper in a 3.5% NaCl solution. The experimental investigations suggest that the number of heteroatoms present in the GG-AMPS coating helps with the good binding to the copper surfaces, thereby reducing the corrosion rate. The SECM suggests that the inductive effect is because of the GG-AMPS coating. SEM and AFM suggest the potential corrosion mitigation of GG-AMPS coatings on copper surfaces in a 3.5% NaCl solution.

Author Contributions: Conceptualization, A.S. and M.L.; methodology, M.L. and E.I.; software, A.S.; validation, A.S. and Y.L.; formal analysis and investigation, A.S. and M.L.; resources, data curation and writing—original draft preparation, Y.L., A.S., M.L. and E.I.; writing—review and editing, visualization and supervision, A.S.; project administration and funding acquisition, A.S., Y.L. and E.I. All authors have read and agreed to the published version of the manuscript.

Funding: The authors are thankful to the Sichuan 1000 Talent Fund; the financial assistance provided by the Youth Scientific and Innovation Research Team for Advanced Surface Functional Materials, Southwest Petroleum University (No. 2018CXTD06); and the open fund project (No. X151517KCL42).

Acknowledgments: Authors would like to acknowledge the help provided by Mumtaz Ahmad Quraishi and Kashif Rahmani Ansari, KFUPM, Saudi Arabia. Authors also extend gratitude to Xu Xihua for her help in the experimental procedures.

Conflicts of Interest: The authors declare no conflict of interest.

References

1. Singh, A.; Ituen, E.; Ansari, K.R.; Chauhan, D.S.; Quraishi, M.A. Surface protection of X80 steel by Epimedium extract and its iodide-modified composites in simulated acid wash solution: A greener approach for corrosion inhibition. *New J. Chem.* **2019**, *43*, 8527–8538. [CrossRef]
2. Du, P.; Li, J.; Zhao, Y.; Dai, Y.; Yang, Z.; Tian, Y. Corrosion characteristics of Al alloy/galvanized-steel Couple in NaCl solution. *Int. J. Electrochem. Sci.* **2018**, *13*, 11164–11179. [CrossRef]
3. Antonijevic, M.M.; Petrovic, M.B. Copper corrosion inhibitors. A review. *Int. J. Electrochem. Sci.* **2008**, *3*, 1–28.
4. Ramachandran, S.; Jovancicevic, V. Molecular modeling of the inhibition of mild copper carbon dioxide corrosion by imidazolines. *Corrosion* **1999**, *55*, 259–267. [CrossRef]
5. Ansari, K.R.; Quraishi, M.A.; Singh, A. Pyridine derivatives as corrosion inhibitors for N80 in 15% HCl: Electrochemical, surface and quantum chemical studies. *Measurement* **2015**, *76*, 136–147. [CrossRef]
6. Sastri, V.S. Green Corrosion Inhibitors. In *Theory and Practice*, 1st ed.; John Wiley & Sons: Hoboken, NJ, USA, 2011.
7. Yang, Y.; Yin, C.; Singh, A.; Lin, Y. Electrochemical study of commercial and synthesized green corrosion inhibitors for N80 steel in acidic liquid. *New J. Chem.* **2019**, *43*, 16058–16070.
8. Fonseca, T.; Gigante, B.; Gilchrist, T.L. A short synthesis of phenanthro [2,3-d] imidazoles from dehydroabietic acid. Application of the methodology as a convenient route to benzimidazoles. *Tetrahedron* **2001**, *57*, 1793–1799. [CrossRef]
9. Jevremovic', I.; Singer, M.; Nešic', S.; Miškovic'-Stankovic', V. Inhibition properties of self-assembled corrosion inhibitor talloil diethylenetriamine imidazoline for mild copper corrosion in chloride solution saturated with carbon dioxide. *Corros. Sci.* **2013**, *77*, 265–272. [CrossRef]
10. Singh, A.; Ansari, K.R.; Quraishi, M.A.; Lgaz, H. Effect of electron donating functional groups on corrosion inhibition of J55 steel in sweet corrosive environment: Experimental, density functional theory and molecular dynamic simulation. *Materials* **2019**, *12*, 17. [CrossRef]
11. Desimone, M.P.; Gordillo, G.; Simison, S.N. The effect of temperature and concentration on the corrosion inhibition mechanism of an amphiphilic amidoamine in CO_2 saturated solution. *Corros. Sci.* **2011**, *53*, e4033–e4043. [CrossRef]
12. Gonzalez-Rodriguez, J.G.; Zeferino-Rodriguez, T.; Ortega, D.M.; Serna, S.; Campillo, B.; Casales, M.; Valenzuela, E.; Juarez-Islas, J. Effect of microstructure on the CO_2 corrosion inhibition by carboxy amidoimidazolines on a pipeline copper. *Int. J. Electrochem. Sci.* **2017**, *2*, 883–896.
13. He, Y.; Yang, R.; Zhou, Y.; Ma, L.; Zhang, L.; Chen, Z. Water soluble Thiosemicarbazideimidazole derivative as an efficient inhibitor protecting P110 carbon copper from CO_2 corrosion. *Anti Corros. Methods Mater.* **2016**, *63*, 437–444. [CrossRef]
14. He, Y.; Zhou, Y.; Yang, R.; Ma, L.; Chen, Z. Imidazoline derivative with four imidazole reaction centers as an efficient corrosion inhibitor for anti-CO_2 corrosion. *Russ. J. Appl. Chem.* **2015**, *88*, 1192–1200. [CrossRef]
15. Kandemirli, F.; Sagdinc, S. Theoretical study of corrosion inhibition of amides and thiosemicarbazones. *Corros. Sci.* **2007**, *49*, 2118–2130. [CrossRef]
16. Biswas, A.; Das, D.; Lgaz, H.; Pal, S.; Nair, U.G. Biopolymer dextrin and poly (vinyl acetate) based graft copolymer as an efficient corrosion inhibitor for mild steel in hydrochloric acid: Electrochemical, surface morphological and theoretical studies. *J. Mol. Liq.* **2019**, *275*, 867–878. [CrossRef]
17. Biswas, A.; Pal, S.; Udayabhanu, G. Experimental and theoretical studies of xanthan gum and its graft co-polymer as corrosion inhibitor for mild steel in 15% HCl. *Appl. Surf. Sci.* **2015**, *30*, 173–183. [CrossRef]
18. Biswas, A.; Mourya, P.; Mondal, D.; Pal, S.; Udayabhanu, G. Grafting effect of gum acacia on mild steel corrosion in acidic medium: Gravimetric and electrochemical study. *J. Mol. Liq.* **2018**, *251*, 867–878. [CrossRef]
19. *ASTM G31-72 Standard Practice for Laboratory Immersion Corrosion Testing of Metals*; ASTM International: West Conshohocken, PA, USA, 2004.
20. Singh, A.; Ansari, K.R.; Haque, J.; Dohare, P.; Lgaz, H.; Salghi, R.; Quraishi, M.A. Effect of electron donating functional groups on corrosion inhibition of mild steel in hydrochloric acid: Experimental and quantum chemical study. *J. Taiwan Inst. Chem. Eng.* **2018**, *82*, 470–479. [CrossRef]

21. Singh, A.; Soni, N.; Deyuan, Y.; Kumar, A. A combined electrochemical and theoretical analysis of environmentally benign polymer for corrosion protection of N80 steel in sweet corrosive environment. *Results Phys.* **2019**, *13*, 102116. [CrossRef]

22. Xu, X.; Singh, A.; Sun, Z.; Ansari, K.R.; Lin, Y. Electrochemical, surface and quantum chemical studies of novel imidazole derivatives as corrosion inhibitors for J55 steel in sweet corrosive environment. *R. Soc. Open Sci.* **2017**, *4*, 170933–170951. [CrossRef]

23. Ansari, K.R.; Quraishi, M.A. Experimental and computational studies of naphthyridine derivatives as corrosion inhibitor for Copper in 15% hydrochloric acid. *Physica E* **2015**, *69*, 322–331. [CrossRef]

24. Ansari, K.R.; Quraishi, M.A.; Singh, A.; Ramkumar, S.; Obot, I.B. Corrosion inhibition of Copper in 15% HCl by pyrazolone derivatives: Electrochemical, surface and quantum chemical studies. *RSC Adv.* **2016**, *6*, 24130–24141. [CrossRef]

25. Li, X.H.; Deng, S.D.; Fu, H.; Mu, G.N. Inhibition by tween-85 of the corrosion of cold rolled copper in 1.0 M hydrochloric acid solution. *J. Appl. Electrochem.* **2009**, *39*, 1125–1135. [CrossRef]

26. Singh, A.; Ansari, K.R.; Quraishi, M.A.; Lgaz, H.; Lin, Y. Synthesis and investigation of pyran derivatives as acidizing corrosion inhibitors for Copper in hydrochloric acid: Theoretical and experimental approaches. *J. Alloys Compd.* **2018**, *762*, 347–362. [CrossRef]

27. Ansari, K.R.; Quraishi, M.A.; Singh, A. Schiff's base of pyridyl substituted triazoles as new and effective corrosion inhibitors for mild copper in hydrochloric acid solution. *Corros. Sci.* **2014**, *79*, 5–15. [CrossRef]

28. Haque, J.; Ansari, K.R.; Srivastava, V.; Quraishi, M.A.; Obot, I.B. Pyrimidine derivatives as novel acidizing corrosion inhibitors for Copper useful for petroleum industry: A combined experimental and theoretical approach. *J. Ind. Eng. Chem.* **2017**, *49*, 176–188. [CrossRef]

29. Singh, A.; Ebenso, E.E.; Quraishi, M.A.; Lin, Y. 5, 10, 15, 20-Tetra (4-pyridyl)-21H, 23H-porphine as an effective corrosion inhibitor for Copper in 3.5% NaCl solution. *Int. J. Electrochem. Sci.* **2014**, *9*, 7495–7505.

30. Ansari, K.R.; Quraishi, M.A.; Singh, A. Isatin derivatives as a non-toxic corrosion inhibitor for mild copper in 20% H_2SO_4. *Corros. Sci.* **2015**, *95*, 62–70. [CrossRef]

31. Singh, A.; Ansari, K.R.; Kumar, A.; Liu, W.; Songsong, C.; Lin, Y. Electrochemical, surface and quantum chemical studies of novel imidazole derivatives as corrosion inhibitors for J55 copper in sweet corrosive environment. *J. Alloys Compd.* **2017**, *712*, 121–133. [CrossRef]

32. Singh, A.; Ansari, K.R.; Xu, X.; Sun, Z.; Kumar, A.; Lin, Y. An impending inhibitor useful for the oil and gas production industry: Weight loss, electrochemical, surface and quantum chemical calculation. *Sci. Rep.* **2017**, *7*, 14904–14921. [CrossRef]

33. Singh, A.; lin, Y.; Obot, I.B.; Ebenso, E.E. Macrocyclic inhibitor for corrosion of N80 steel in 3.5% NaCl solution saturated with CO_2. *J. Mol. Liq.* **2016**, *219*, 865–874. [CrossRef]

34. Singh, A.; Ansari, K.R.; Chauhan, D.S.; Quraishi, M.A.; Lgaz, H.; Chung, I.M. Comprehensive investigation of steel corrosion inhibition at macro/micro level by ecofriendly green corrosion inhibitor in 15% HCl medium. *J. Colloid Interface Sci.* **2020**, *560*, 225–236. [CrossRef] [PubMed]

 *coatings*

Article

Polyethylene Glycol (PEG) Modified Porous Ca$_5$(PO$_4$)$_2$SiO$_4$ Bioceramics: Structural, Morphologic and Bioactivity Analysis

Pawan Kumar [1,*], Meenu Saini [1,*], Vinod Kumar [2], Brijnandan S. Dehiya [1], Anil Sindhu [3], H. Fouad [4,5,*], Naushad Ahmad [6], Amer Mahmood [7] and Mohamed Hashem [8]

1 Department of Materials Science and Nanotechnology, Deenbandhu Chhotu Ram University of Science and Technology, Murthal 131039, India; drbrijdehiya.msn@dcrustm.org

2 Department of Biotechnology, Deenbandhu Chhotu Ram University of Science and Technology, Murthal 131039, India; indoravinod2@gmail.com

3 Department of Bio and Nanotechnology, Guru Jambheshwar University of Science and Technology, Hisar 125001, India; sindhu.anil@gmail.com

4 Applied Medical Science Department, Community College, King Saud University, P.O. Box 10219, Riyadh 11433, Saudi Arabia

5 Biomedical Engineering Department, Faculty of Engineering, Helwan University, P.O. Box, Helwan 11792, Egypt

6 Chemistry Department, College of Science King Saud University, Riyadh 11451, Saudi Arabia; exactlykot@gmail.com

7 Stem Cell Unit, Department of Anatomy, College of Medicine, King Saud University, Riyadh 11461, Saudi Arabia; amer_dk@yahoo.com

8 Dental Health Department, College of Applied Medical Sciences, King Saud University, P.O Box 10219, Riyadh 11433, Saudi Arabia; mihashem@ksu.edu.sa

* Correspondence: pawankamiya@yahoo.in (P.K.); meenu.rschmsn@dcrustm.org (M.S.); menhfef@ksu.edu.sa (H.F.)

Received: 10 March 2020; Accepted: 29 May 2020; Published: 31 May 2020

Abstract: Bioceramics are class of biomaterials that are specially developed for application in tissue engineering and regenerative medicines. Sol-gel method used for producing bioactive and reactive bioceramic materials more than those synthesized by traditional methods. In the present research study, the effect of polyethylene glycol (PEG) on Ca$_5$(PO$_4$)$_2$SiO$_4$ (CPS) bioceramics was investigated. The addition of 5% and 10% PEG significantly affected the porosity and bioactivity of sol-gel derived Ca$_5$(PO$_4$)$_2$SiO$_4$. The morphology and physicochemical properties of pure and modified materials were evaluated using scanning electron microscopy (SEM), X-ray powder diffraction (XRD), transmission electron microscopy (TEM) and Fourier-transform infrared spectroscopy (FTIR), respectively. The effect of PEG on the surface area and porosity of Ca$_5$(PO$_4$)$_2$SiO$_4$ was measured by Brunauer–Emmett–Teller (BET). The results obtained from XRD and FTIR studies confirmed the interactions between PEG and CPS. Due to the high concentration of PEG, the CPS-3 sample showed the largest-sized particle with an average of 200.53 μm. The porous structure of CPS-2 and CPS-3 revealed that they have a better ability to generate an appetite layer on the surface of the sample when immersed in simulated body fluid (SBF) for seven days. The generation of appetite layer showed the bioactive nature of CPS which makes it a suitable material for hard tissue engineering applications. The results have shown that the PEG-modified porous CPS could be a more effective material for drug delivery, implant coatings and other tissue engineering applications. The aim of this research work is to fabricate SBF treated and porous polyethylene glycol-modified Ca$_5$(PO$_4$)$_2$SiO$_4$ material. SBF treatment and porosity of material can provide a very useful target for bioactivity and drug delivery applications in the future.

Keywords: calcium phosphate silicate; PEG; bioceramics; sol-gel preparation; hard tissue engineering

1. Introduction

$Ca_5(PO_4)_2SiO_4$ bio-ceramic is a fully loaded compound consisting of Ca, P and Si elements [1]. The calcium–phosphate–silicate (CPS) ceramics are considered as biocompatible materials which is exclusively utilized as implantable materials for bone defects repairing [2]. Silicon-based bioceramics permit to form a functional silanol group with Si–O–H connectivity on the material surface [3]. The silanol group induces the formation of bone-like apatite by attracting Ca^{2+} and PO_4^{3-} through an ion-exchange process in an artificial solution similar to blood plasma [4,5]. This apatite is broadly utilized biomaterial to repair and reconstruct hard tissue defects [6,7]. Thus, researches on silicon-containing bioceramics have received a considerable attention from the biomaterials scientists as such biomaterials are considered as bone substitute materials [8,9]. However, various drawbacks such as less mechanism properties, low chemical stability and reduced bioactivity, the clinical applications of CPS bioceramics are limited [10,11]. It was researched that for a reliable bone fixation, the porosity of implant material should be over 70% so that the body fluids can penetrate easily for the better bone growth [12,13]. In order to get a bioactive and porous material, sol-gel derived polyethylene glycol-modified CPS was prepared in which polyethylene glycol (PEG) was used to modify the phase of $Ca_5(PO_4)_2SiO_4$ via crosslinking of PEG diacrylate chain with PO_4^{3-} and Ca^{2+} probably by the OH^- exchange reactions [14,15]. PEG has been widely accepted as a phase change material, which can be altered through its congruent melting behavior and low vapor pressure [16]. Low molecular weight polyethylene glycol (PEG) is a highly non-immunogenic, non-toxic, biocompatible and biodegradable polymer [17,18]. It possesses a straight poly-ether diol chain that has hydroxyl groups and also shows covalent binding with proteins, phospholipids, functional groups, fluorescent probes, etc [19]. PEG will be a promising agent that may enhance the biocompatibility of $Ca_5(PO_4)_2SiO_4$ by generating porosity in the material. The use of larger PEG chains resulted in more agglomerated hollow particles [20]. Sol-gel process is a flexible and favorable method due to its low-temperature, high purity, easy doping and cost effective approach to prepare various nanocomposites and other nanobiomaterials [21,22]. The reaction time may affect the particle size and stability in the sol-gel process [23]. During the reaction monomers converted into colloidal solution (sol) and then into gel or integrated network of discrete particles [24,25]. Due to bioactive properties, the silica-based glass networks are prepared in various shapes, sizes and hence applied for variety of biomedical applications [26]. Sol-gel synthesized bioceramics are highly biocompatible with controlled degradation rate and effortlessly metabolized in the body [27,28].

In the current research work, we synthesized a pure phase of $Ca_5(PO_4)_2SiO_4$ and PEG-modified porous $Ca_5(PO_4)_2SiO_4$ bioceramic materials through the sol-gel method and investigated various properties. The highly bioactive and porous PEG-modified CPS can be utilized for tissue engineering and drug delivery.

2. Materials and Methods

2.1. Synthesis of Porous Calcium Phosphate Silicate

Sol-gel synthesis was convenient because it permits direct fabrication of bioceramics with different configurations. The synthesis of PEG-modified $Ca_5(PO_4)_2SiO_4$ was carried out as follows: 12.17 mL tetraethyl orthosilicate (TEOS, ≥98% Sigma Aldrich, Saint Louis, MO, USA), 150 mL ethanol, 0.80 g P_2O_5 (≥99.9% Sigma Aldrich) and 6.68 g $Ca(NO_3)_2 \cdot 4H_2O$ (≥99.9% Sigma Aldrich) was mixed stepwise to get 680 mL homogenous mixture. To make composite, 5% and 10% *w/v* of PEG 400 were dispersed in 100 mL distilled water and mix with the above mentioned homogenous mixture. In acidic conditions, TEOS completely hydrolyzed and obtained $Si(OH)_4$ which slow down the condensation rate [22]. Add ammonia (25%) solution to make pH 11 of the mixture which boosts the rate of the gelation process, ideal for the formation of smaller aggregates. TEOS works as a principal network forming agent during gelation. The change in synthesis conditions or parameters such as pH, temperature and additives affect the silica-based glass networks which produce various shapes, sizes and formats products [26]. The different steps used for the synthesis of PEG-modified $Ca_5(PO_4)_2SiO_4$ are shown through the schematic diagram, see Figure 1. Naming of the samples was done on the basis of concentration variation of additive, i.e., PEG in $Ca_5(PO_4)_2SiO_4$; for pure $Ca_5(PO_4)_2SiO_4$ without PEG was assigned to CPS-1. Similarly, for PEG 5% ND 10% by weight in $Ca_5(PO_4)_2SiO_4$ was assigned CPS-2 and CPS-3, respectively. Further, these three samples were used for different characterization.

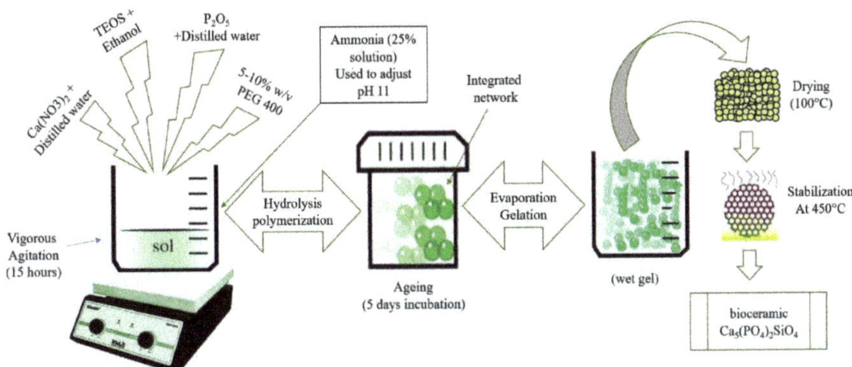

Figure 1. Synthesis of porous calcium phosphate silicate.

2.2. Characterizations of Porous Calcium Phosphate Silicate

The phase identification of the synthesized materials was investigated using XRD (Rigaku Ultima IV, Tokyo, Japan). The measurement was taken at 45 kV voltage and 40 mA anodic current. XRD patterns were acquired at a diffraction angle from 15° to 60°. The compositional information of the prepared samples was investigated by FTIR (Perkin Elmer Frontier FTIR, MA, USA). First, the obtained powder was mixed with KBr in an appropriate ratio and then after applying pressure. The mixture was converted into pellets. For investigations of obtained spectra, the background spectra were calibrated with KBr. The morphology and elemental composition of the samples was examined using SEM (JEOL, JSM 6100, Akishima, Tokyo, Japan) and by energy dispersive spectroscopy (EDS) attached with same, respectively. On sputter was then to sputter coat the samples with a palladium layer. After 30 nm palladium coating, observations were done at an accelerating voltage of 20 kV and 10 Pa. The powerful size of the pore was figured as the mean distances across of sample pores. The nanoparticles synthesis confirmed through TEM (TECNAI 200 kV, Hillsboro, OR, USA) at SAIF in AIIMS, New Delhi. To provide contrast under magnification, nanoparticles were suspended in water (1 mg/mL), placed on copper grids of 0.037-mm size and then stained with a 2 g/100 mL uranyl acetate aqueous stain. Before viewing under 50,000 to 120,000 times magnification, surplus liquid on Mesh was wiped off with filter study and the grid was allowed to air dry. Observations were performed at 80 kV. Brunauer–Emmett–Teller (BET) (BELSORP mini II, Osaka, Japan) technique was used to measure the porosity and surface area of the prepared materials. Tris-HCl-buffered synthetic body fluid (SBF) was used to check bioactivity of the sample after 7 days at 37 °C in an incubator. The thermal stability of the prepared material was evaluated by the thermogravimetric analysis (TGA; Perkin Elmer STA 6000, Waltham, MA, USA) operated under nitrogen flow in the temperature range from 50 to 800 °C at 10 °C/min. Using TGA, by inducing heat to the sample, the chemical reactions and physical changes due to dehydration, decomposition and oxidation can be evaluated.

3. Results and Discussion

3.1. Physiochemical Analysis of Calcium Phosphate Silicate Materials

The sol-gel derived white powdered samples of CPS and PEG-modified CPS were heated at 450 °C to get phase or structural transformation. In the XRD spectra, the distinct peaks of the pure phase of CPS-1 were identified and matched with the standard database card number 00-901-1950 and PDF 40-0393 [29].

The XRD patterns of the PEG-modified CPS bioceramics showed modification in peaks [30]. The addition of 5% and 10% PEG in CPS make some changes in the sample, generate and demolish several phases or peaks in CPS-2 and CPS-3 (Figure 2A). The heat treatment (450 °C) to CPS-2 (5% PEG) and CPS-3 (10% PEG) removed the precursor residues of PEG, limiting the densification of material [31]. During the heat treatment, PEG started to decompose within a temperature range of around 250–300 °C; this can completely remove PEG from the sample [32,33]. The removal/degradation of PEG may generate a porous structure due to structural rearrangement by various chemical reactions, based on several treatment temperature and duration. The sintered sample of CPS-2 (5% PEG) and CPS-3 (10% PEG) also confirmed the presence of wollastonite [PDF 50–0905].

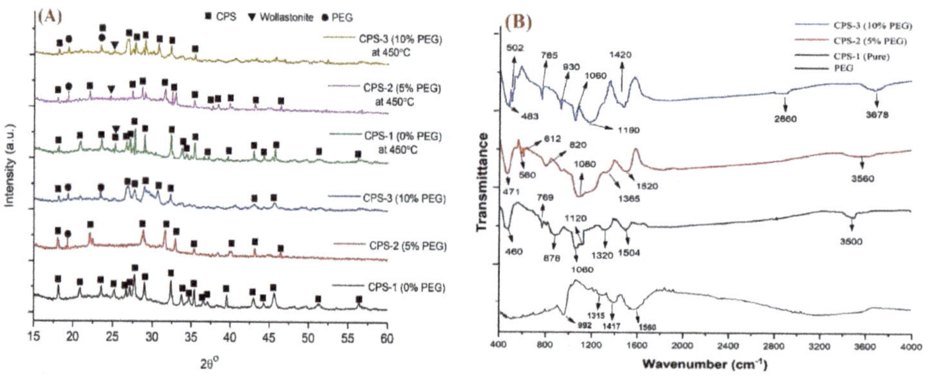

Figure 2. Typical (**A**) XRD pattern and (**B**) FTIR spectra of porous calcium phosphate silicate.

FTIR (Figure 2B) spectrum data revealed O–H stretching vibrations observed at 3500, 3560 and 3678 cm^{-1} [34,35] and O–H deformation at 765 and 769 cm^{-1}. The band occurred at 2860 cm^{-1} due to the existence of the C–H group stretching in CPS-3 as a result of the presence of PEG [36]. The intense bands within 450–510 cm^{-1} correspond to Si–O–, P–O, PO_4^{3-} and SiO_4^{4-}, while the absorption bands at 1060 cm^{-1} assigned to the vibration of the Si–O–Si [30,34]. The functional group SiO_4^{4-} was also recognized by nearly at 612 and 878 cm^{-1} [30]. The absorption bands at 820 and 878 cm^{-1} represent Si–CH$_3$ and SiO_4^{4-} functional groups. The band between 1000–1200 cm^{-1} is associated with the Si–O– stretching, while the band at 930 cm^{-1} corresponds to Si–O– with one non-bridging oxygen [34]. The appearance of a medium stretching vibration at 1320 cm^{-1} was because of the C=O group. The pure PEG sample showed bands at 992, 1315, 1417 and 1560 cm^{-1} while PEG-modified sample showed bands at 1320 and 1365 cm^{-1}, denoting CH$_3$ and C–O stretching vibrations.

3.2. Morphologic Characterizations of Calcium Phosphate Silicate Materials

The high-resolution 2 D TEM images revealed the particle size estimation of PEG-modified CPS-1 (Figure 3A,B), CPS-2 (Figure 3C,D) and CPS-3 (Figure 3E,F) samples. Figure 3A shows a 506-nm-sized crystalline particle of CPS-1. CPS-2 (Figure 3C) revealed 265.4-nm-sized irregular particles while CPS-3 has not revealed any proper shape. Crystallite size is calculated using the Scherrer formula from XRD patterns for all synthesized samples. The calculated crystal size of CPS-1, CPS-2 and CPS-3 is 24,

18 and 15 nm, respectively. These calculated results are quite different from TEM results. Because the crystallite size determined using Scherrer formula from XRD patterns provided an average value of the bulk sample since the diffraction occurs from a considerable volume of the sample. Apart from that in TEM, we found the crystallite size from a very local area that may not be the representative size of the bulk sample. Removal of PEG at high temperatures during heat treatment may lead to a highly ordered porous structure, as observed in CPS-2 and CPS-3 (Figure 3). At the 200-nm-scale, all the samples showed significant structural differences; CPS-1 (0% PEG) looked like a dense material, while CPS-3 (10% PEG) displayed a better porous structure than the CPS-2 (5% PEG) sample.

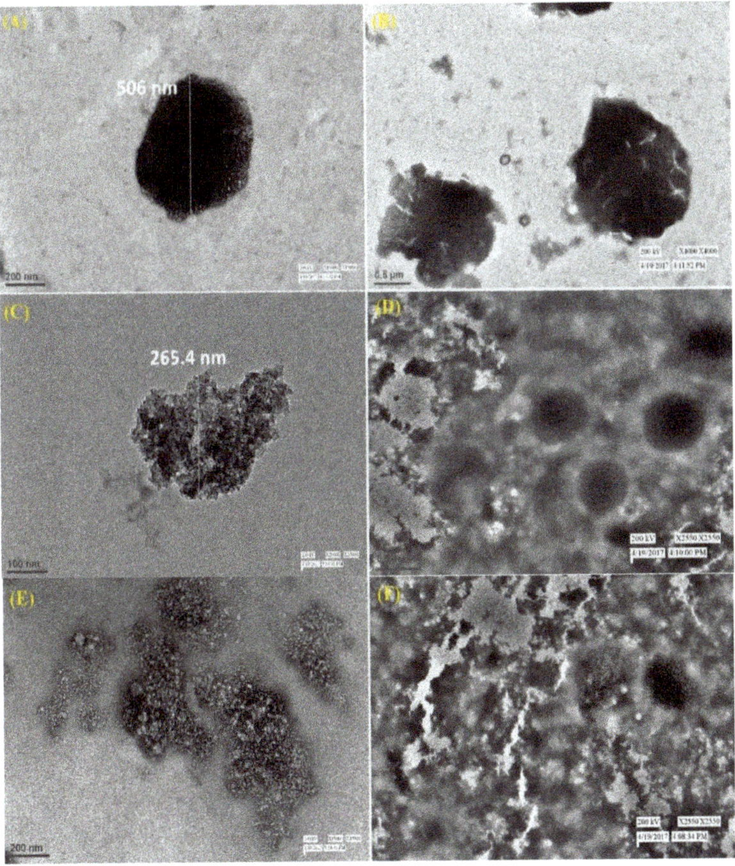

Figure 3. Typical transmission electron microscopy (TEM) analysis of Ca5(PO4)2SiO4 (CPS)-1 (**A,B**), CPS-2 (**C,D**) and CPS-3 (**E,F**).

The increment of PEG concentration (more than 10% *w/v*) led to the densification of materials, which reduced the porosity of the CPS. The SEM results revealed the irregular micro size crystalline particles of CPS-1 (Figure 4A) and CPS-2 (Figure 4B) revealed amorphous particles. The SEM image of CPS-3 showed a porous microstructure, observed after the removal of PEG after heat treatment, see Figure 4C. The morphology of this specimen significantly display that will be beneficial for future applications, for instance tissue engineering and drug loading. The optimized favorite sample CPS-3 was used investigated through EDS analysis (Figure 4D) which showed the relative concentration of the Si, Ca, P and C elements in the synthesized sample. The results obtained from EDS analysis depend on several factors, to name a few, sample topography, beam parameters, field noises (electronic and external fields), acquisition settings, detector type and atomic number of the elements. From particle size distribution histogram (Figure 4E), it is concluded that particle size of CPS-1 is to be under 10 μm, for CPS-2 it is estimated to be under 100 μm, while for CPS-3, it observed to be under 160 μm due to agglomeration of particles.

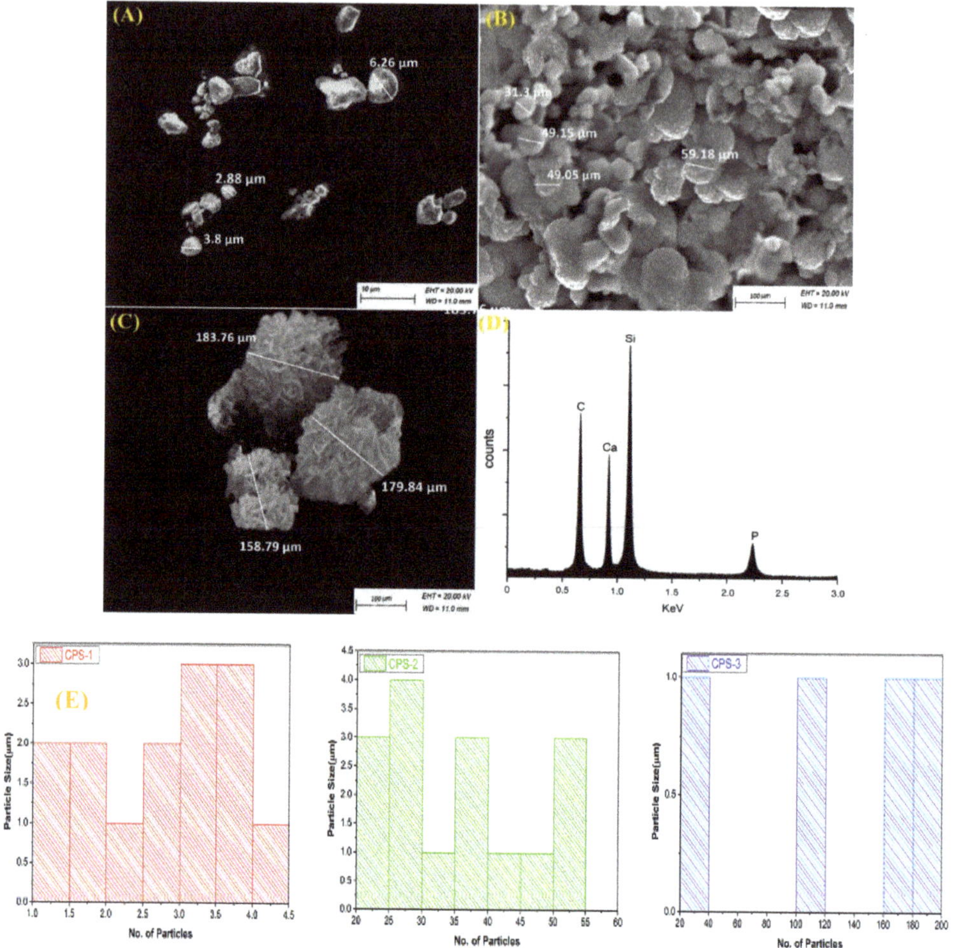

Figure 4. Typical SEM analysis of CPS-1 (**A**), CPS-2 (**B**), CPS-3 (**C**), energy dispersive spectroscopy (EDS) spectrum of CPS-3 (**D**) and particle size distribution histogram for CPS-1, CPS-2, CPS-3 (**E**).

3.3. Porosity Measurements for Calcium Phosphate Silicate Materials

The Brunauer, Emmett and Teller (BET) method is a promising technique to determine the surface area through physical adsorption–desorption isotherm analysis [37]. The porosity of a material depends upon the concentration and type or nature of the template used during the fabrication process [38]. The reparative bone formation and inflammatory response are also influenced by the morphologies of the materials [39,40]. From BET analysis (Figure 5A,B), CPS-3 revealed a surface area of 30.6672 $m^2 \cdot g^{-1}$, pore volume of 0.9722 $cm^3 \cdot g^{-1}$ and pore diameter of 4.58 nm, while CPS-2 revealed a surface area of 27.9840 $m^2 \cdot g^{-1}$, pore volume of 0.6243 $cm^3 \cdot g^{-1}$ and pore diameter of 2.84 nm. The addition of PEG affects the microstructure as well as the porosity of the CPS. The BET result revealed that the

CPS-3 (10% PEG) has shown better porosity that will be a key factor in application part. At 450 °C, the decomposition of PEG generated the porous network. The bio factors include genes or cells, proteins and nutrients can easily exchange through the porous structure of materials. Porosity and pore size possess high impact on the application of the material [3,41].

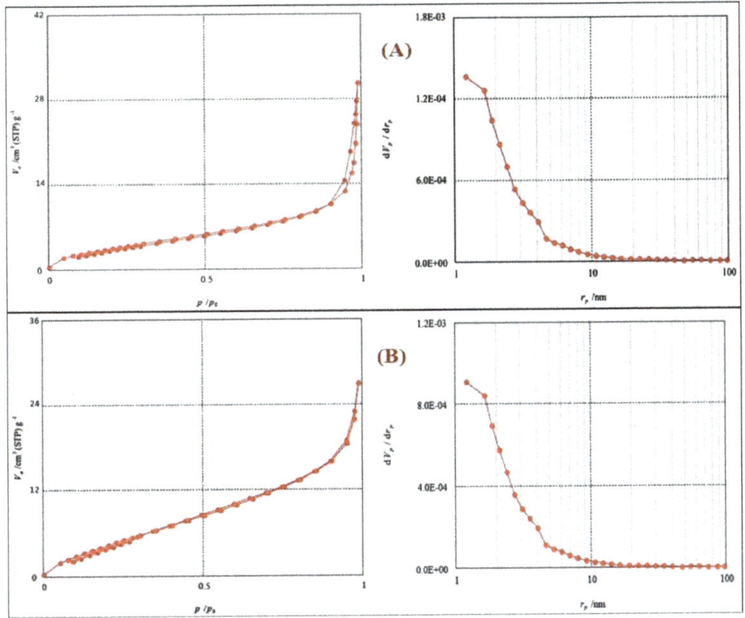

Figure 5. Adsorption–desorption isotherm and Barrett, Joyner, and Halenda (BJH) plot of CPS-3 (**A**) and CPS-2 (**B**).

3.4. Bioactivity Analysis

SBF provides the same environment as blood plasma, where surface dissolution starts mineralization through the slow release of ions in the solvent [33]. When materials were immersed in the SBF solution, a set of reactions such as ion exchange, precipitation and dissolution occurred for the apatite formation [42]. The SBF treatment generated two new groups, CO_3^{2-} and PO_4^{3-}, at 1420 and 1480 cm^{-1}, which are due to the absorbance of CO_2 (Figure 6D). The generation of new compounds are directed by both the immersion parameters (immersion time, temperature and pH, temperature) and surface characteristics of the materials [43]. The presence of CO_3^{2-} and PO_4^{3-} groups are associated with apatite formation on the surface of the sample after a seven-day immersion [30,44]. Further, CO_3^{2-} and PO_4^{3-} ions contribute to nucleation and subsequent surface mineralization that leads to actual apatite formation [45,46]. The in vitro apatite-forming ability of material often successfully predicts the actual bioactivity of biomaterials [47]. The bands at 1653 and 1470 cm^{-1} denoted CO_3^{2-}, while peaks at 1314 and 2790 cm^{-1} were attributed to C–H group. The peaks appearing at 561 and 880 cm^{-1} are due to bending mode of O–P–O and variable symmetry of HPO_4^{3-}, respectively [48]. The band at 1025 cm^{-1} was assigned to the presence of the PO_4^{3-} group [49,50]. The surface of the material released Ca^{2+}, HPO_4^{2-} and PO_4^{3-} ions and absorbed calcium and phosphate ions from SBF. The incorporation of other electrolytes, such as CO_3^{2-} and Mg^{2+} ions, started to generate the apatite layer [51]. The mineralization behavior of CPS-1, CPS-2 and CPS-3 was shown in Figure 6A–C and also confirmed through FTIR results, as shown in Figure 6D.

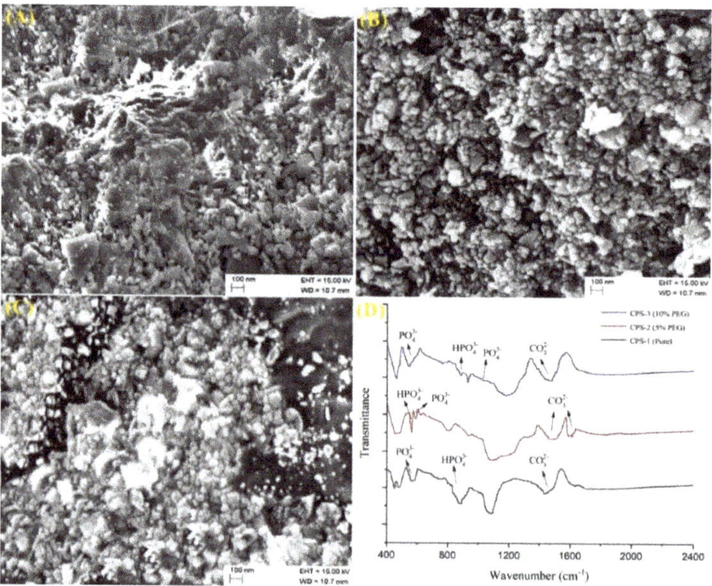

Figure 6. SEM analysis of synthetic body fluid (SBF)-treated CPS-1 (**A**), CPS-2 (**B**), CPS-3 (**C**) and FTIR analysis (**D**).

3.5. Thermal Gravimetric Analysis

The physical and morphologic transformations of the PEG-modified CPS samples analyzed through thermal gravimetric analysis. It is well known that the chemical structures are altered by heat treatment as it leads to the thermal decomposition of the materials. TGA results presented in Figure 7, weight loss in CPS-1 and CPS-2 divided into three main steps (S-1, S-2 and S-3): removal of –OH groups, polymeric phase (PEG) and burnout of the CPS mass (Ca, P and Si) while CPS-3 revealed weight loss in two stages. In all the samples, the initial weight loss was confirmed because of the release of absorbed moisture contents. CPS-2 and CPS-3 showed more weight loss than CPS-1 that was because of the PEG thermal degradation within a temperature range of around 250–300 °C. CPS-3 fabricated with 10% PEG but thermal decomposition of the organic groups, generated porous microstructure and microporous materials showed large specific surface that support more degradation at high temperature.

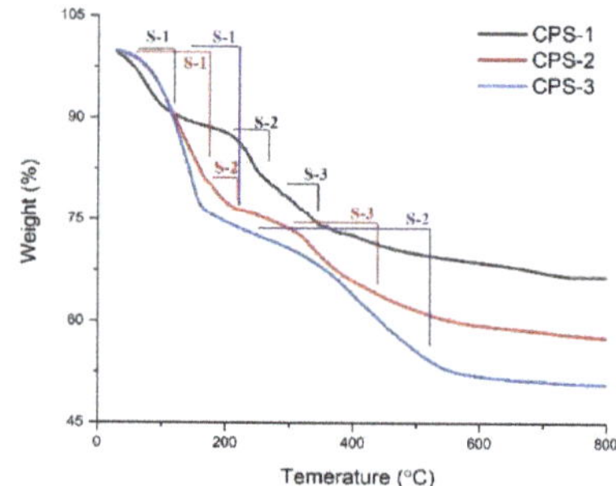

Figure 7. Thermos-gravimetric analysis (TGA) analysis of CPS-1, CPS-2 and CPS-3.

4. Conclusions

In summary, the effect of PEG on various properties of CPS was investigated. The use of PEG improves the morphology, physiology and bioactivity of CPS. Porosity and bioactivity of sol-gel-derived samples were greatly influenced by varying the concentration of PEG. The heat treatment at 450 °C plays an important role in the phase modification and porosity generation. This could be attributed, as the concentration of PEG increased, the densification and agglomeration in particles were observed. The formation of the apatite layer on the surface of SBF treated CPS exposed mineralization. PEG-modified $Ca_5(PO_4)_2SiO_4$ demonstrated better in vitro bioactivity than pure CPS, by tempting bone-like apatite in the artificial salt solution SBF. PEG-modified $Ca_5(PO_4)_2SiO_4$ bioceramic (CPS-3) is different from those in conventional material and may be a promising material for implant coatings, drug loading and bone regeneration applications. Future works should determine the optimum concentration for controlled porosity and their applications in soft as well as hard tissue engineering.

Author Contributions: Conception and design of the experiments, P.K., M.S., V.K., B.S.D., A.S., H.F., N.A., A.M. and M.H. implementation of the experiments, analysis of the data, the contribution of the analysis tools and writing the study. All authors have read and agreed to the published version of the manuscript.

Funding: The authors would like to extend their sincere appreciation to the Deanship of Scientific Research at King Saud University for funding this research group (No. RGP-1435-052).

Acknowledgments: The authors would like to extend their sincere appreciation to the Deanship of Scientific Research at King Saud University, Kingdom of Saudi Arabia for support.

Conflicts of Interest: The authors declare no conflict of interest.

References

1. Wu, C.; Chang, J. A review of bioactive silicate ceramics. *Biomed. Mater.* **2013**, *8*, 32001. [CrossRef]
2. Duan, W.; Ning, C.; Tang, T. Cytocompatibility and osteogenic activity of a novel calcium phosphate silicate bioceramic: Silicocarnotite. *J. Biomed. Mater. Res. Part A* **2013**, *101A*, 1955–1961. [CrossRef] [PubMed]
3. Bordbar-Khiabani, A.; Yarmand, B.; Mozafari, M. Emerging magnesium-based biomaterials for orthopedic implantation. *Emerg. Mater. Res.* **2019**, *8*, 305–319. [CrossRef]
4. Lu, W.; Duan, W.; Guo, Y.; Ning, C. Mechanical properties and in vitro bioactivity of $Ca_5(PO_4)_2SiO_4$ bioceramic. *J. Biomater. Appl.* **2012**, *26*, 637–650. [CrossRef] [PubMed]
5. Kumar, P.; Dehiya, B.S.; Sindhu, A. Comparative study of chitosan and chitosan-gelatin scaffold for tissue engineering. *Int. Nano Lett.* **2017**, *7*, 285–290. [CrossRef]
6. Ji, D.D.; Xu, J.Y.; Gu, X.F.; Zhao, Q.M.; Gao, A.G. Preparation, characterization and biocompatibility of bioactive coating on titanium by plasma electrolytic oxidation. *Sci. Adv. Mater.* **2019**, *11*, 1411–1415. [CrossRef]
7. Wang, P.; He, H.; Cai, R.; Tao, G.; Yang, M.; Zuo, H.; Umar, A.; Wang, Y. Cross-linking of dialdehyde carboxymethyl cellulose with silk sericin to reinforce sericin film for potential biomedical application. *Carbohydr. Polym.* **2019**, *212*, 403–411. [CrossRef]
8. Li, J.; Wu, L.; Shen, R.; Hou, Y.; Chen, Z.; Cai, M.; Wei, Z.; Chen, X.; Gao, J. Preparation and biocompatibility of corrosion resistant micro-nano dual structures on the surface of TAx pure titanium dental implants. *Sci. Adv. Mater.* **2019**, *11*, 1656–1665. [CrossRef]
9. Kumar, P.; Dehiya, B.S.; Sindhu, A. Bioceramics for hard tissue engineering applications: A review. *Int. J. Appl. Eng. Res.* **2018**, *13*, 2744–2752.
10. Deng, F.; Wang, F.; Liu, Z.; Kou, H.; Cheng, G.; Ning, C. Enhanced mechanical property of $Ca_5(PO_4)_2SiO_4$ bioceramic by a biocompatible sintering aid of zinc oxide. *Ceram. Int.* **2018**, *44*, 18352–18362. [CrossRef]
11. Chaudhary, S.; Umar, A.; Mehta, S. Surface functionalized selenium nanoparticles for biomedical applications. *J. Biomed. Nanotechnol.* **2014**, *10*, 3004–3042. [CrossRef] [PubMed]
12. Shad, A.A.; Ahmad, S.; Ullah, R.; AbdEl-Salam, N.M.; Fouad, H.; Rehman, N.U.; Hussain, H.; Saeed, E. Phytochemical and biological activities of four wild medicinal plants. *Sci. World J.* **2014**, 857363. [CrossRef] [PubMed]

13. Kumar, P. Nano-TiO$_2$ doped chitosan scaffold for the bone tissue engineering applications. *Int. J. Biomater.* **2018**, 6576157. [CrossRef]

14. Elliott, J.E.; Macdonald, M.; Nie, J.; Bowman, C.N. Structure and swelling of poly(acrylic acid) hydrogels: Effect of pH, ionic strength, and dilution on the crosslinked polymer structure. *Polymer (Guildf.)* **2004**, *45*, 1503–1510. [CrossRef]

15. Jones, D.S.; Andrews, G.P.; Gorman, S.P. Characterization of crosslinking effects on the physicochemical and drug diffusional properties of cationic hydrogels designed as bioactive urological biomaterials. *J. Pharm. Pharmacol.* **2005**, *57*, 1251–1259. [CrossRef]

16. Sun, Q.; Yuan, Y.; Zhang, H.; Cao, X.; Sun, L. Thermal properties of polyethylene glycol/carbon microsphere composite as a novel phase change material. *J. Therm. Anal. Calorim.* **2017**, *130*, 1741–1749. [CrossRef]

17. Paul, W.; Sharma, C.P. Polyethylene glycol modified calcium phosphate microspheres facilitate selective adsorption of immunogobulin G from human blood. *Trends Biomater. Artif. Organs* **2013**, *27*, 20–28.

18. Nagar, P.; Goyal, P.; Gupta, A.; Sharma, A.K.; Kumar, P. Synthesis, characterization and evaluation of retinoic acid-polyethylene glycol nanoassembly as efficient drug delivery system. *Nano Struct. Nano Objects* **2018**, *14*, 110–117. [CrossRef]

19. Mohaisen, M.; Yildirim, R.; Yilmaz, M.; Durak, M. Production of functional yogurt drink, apple and orange juice using nano-encapsulated l. brevis within sodium alginate-based biopolymers. *Sci. Adv. Mater.* **2019**, *11*, 1788–1797. [CrossRef]

20. Dave, V.; Gupta, A.; Singh, P.; Gupta, C.; Sadhu, V.; Reddy, K.R. Synthesis and characterization of celecoxib loaded PEGylated liposome nanoparticles for biomedical applications. *Nano Struct. Nano Objects* **2019**, *18*, 100288.

21. Zhang, X.; Zeng, D.; Li, N.; Wen, J.; Jiang, X.; Liu, C.; Li, Y. Functionalized mesoporous bioactive glass scaffolds for enhanced bone tissue regeneration. *Sci. Rep.* **2016**, *6*, 19361. [CrossRef] [PubMed]

22. Kumar, P.; Dehiya, B.S.; Sindhu, A. Synthesis and characterization of nHA-PEG and nBG-PEG scaffolds for hard tissue engineering applications. *Ceram. Int.* **2019**, *45*, 8370–8379. [CrossRef]

23. Ansari, S.G.; Fouad, H.; Shin, H.S.; Ansari, Z.A. Electrochemical enzyme-less urea sensor based on nano tim oxide synthesized by hydrothermal technique. *Chem. Biol. Interact.* **2015**, *242*, 45–49. [CrossRef] [PubMed]

24. Algarni, H.; AlShahrani, I.; Ibrahim, E.H.; Eid, R.A.; Kilany, M.; Ghramh, H.A.; Abdellahi, M.O.; Sayed, M.A.; Yousef, E.S. In-vitro bioactivity of optical glasses containing strontium oxide (SrO). *J. Nanoelectron. Optoelectron.* **2019**, *14*, 1105–1112. [CrossRef]

25. Ansari, F.; Soofivand, F.; Salavati-Niasari, M. Eco-friendly synthesis of cobalt hexaferrite and improvement of photocatalytic activity by preparation of carbonic-based nanocomposites for waste-water treatment. *Compos. Part B Eng.* **2019**, *165*, 500–509. [CrossRef]

26. Ansari, F.; Sobhani, A.; Salavati-Niasari, M. Simple sol-gel synthesis and characterization of new CoTiO$_3$/CoFe$_2$O$_4$ nanocomposite by using liquid glucose, maltose and starch as fuel, capping and reducing agents. *J. Colloid Interface Sci.* **2018**, *514*, 723–732. [CrossRef]

27. Owens, G.J.; Singh, R.K.; Foroutan, F.; Alqaysi, M.; Han, C.-M.; Mahapatra, C.; Kim, H.-W.; Knowles, J.C. Sol-gel based materials for biomedical applications. *Prog. Mater. Sci.* **2016**, *77*, 1–79. [CrossRef]

28. Algarni, H.; Alshahrani, I.; Ibrahim, E.; Eid, R.; Kilany, M.; Ghramh, H.; Sayed, M.; Reben, M.; Yousef, E. Structural, thermal stability and in vivo bioactivity properties of nanobioglasses containing ZnO. *Sci. Adv. Mater.* **2019**, *11*, 925–935. [CrossRef]

29. Rao, S.H.; Harini, B.; Shadamarshan, R.P.K.; Balagangadharan, K.; Selvamurugan, N. Natural and synthetic polymers/bioceramics/bioactive compounds-mediated cell signalling in bone tissue engineering. *Int. J. Biol. Macromol.* **2018**, *110*, 88–96. [CrossRef]

30. Radev, L.; Hristov, V.; Michailova, I.; Samuneva, B. Sol-gel bioactive glass-ceramics Part I: Calcium phosphate silicate/ wollastonite glass-ceramics. *Cent. Eur. J. Chem.* **2009**, *7*, 317–321. [CrossRef]

31. Lombardi, M.; Gremillard, L.; Chevalier, J.; Lefebvre, L.; Cacciotti, I.; Bianco, A.; Montanaro, L. A comparative study between melt-derived and sol-gel synthesized 45S5 bioactive glasses. *Key Eng. Mater.* **2013**, *541*, 15–30. [CrossRef]

32. Matsuda, A.; Matsuno, Y.; Katayama, S.; Tsuno, T.; Tohge, N.; Minami, T. Physical and chemical properties of titania-silica films derived from poly(ethylene glycol)-containing gels. *J. Am. Ceram. Soc.* **1990**, *73*, 2217–2221. [CrossRef]

33. Kumar, P.; Dehiya, B.S.; Sindhu, A. Ibuprofen-loaded CTS/nHA/nBG scaffolds for the applications of hard tissue engineering. *Iran. Biomed. J.* **2019**, *23*, 190–199. [CrossRef] [PubMed]

34. Ouis, M.; Abdelghany, A.; Elbatal, H. Corrosion mechanism and bioactivity of borate glasses analogue to Hench's bioglass. *Process. Appl. Ceram.* **2012**, *6*, 141–149. [CrossRef]

35. Davar, F.; Salavati-Niasari, M.; Mir, N.; Saberyan, K.; Monemzadeh, M.; Ahmadi, E. Thermal decomposition route for synthesis of Mn_3O_4 nanoparticles in presence of a novel precursor. *Polyhedron* **2010**, *29*, 1747–1753. [CrossRef]

36. Kumar, P.; Saini, M.; Dehiya, B.S.; Umar, A.; Sindhu, A.; Mohammed, H.; Al-Hadeethi, Y.; Guo, Z. Fabrication and in-vitro biocompatibility of freeze-dried CTS-nHA and CTS-nBG scaffolds for bone regeneration applications. *Int. J. Biol. Macromol.* **2020**, *149*, 1–10. [CrossRef]

37. Salavati-Niasari, M. Ship-in-a-bottle synthesis, characterization and catalytic oxidation of styrene by host (nanopores of zeolite-Y)/guest ([bis(2-hydroxyanil)acetylacetonato manganese(III)]) nanocomposite materials (HGNM). *Microporous Mesoporous Mater.* **2006**, *95*, 248–256. [CrossRef]

38. Kumar, P.; Saini, M.; Kumar, V.; Singh, M.; Dehiya, B.S.; Umar, A.; Ajmal Khan, M.; Alhuwaymel, T.F. Removal of Cr (VI) from aqueous solution using VO2(B) nanoparticles. *Chem. Phys. Lett.* **2019**, 136934. [CrossRef]

39. Sampath, U.G.T.M.; Ching, Y.C.; Chuah, C.H.; Sabariah, J.J.; Lin, P.-C. Fabrication of porous materials from natural/synthetic biopolymers and their composites. *Materials* **2016**, *9*, 991. [CrossRef]

40. Fathi, M.H.; Hanifi, A.; Mortazavi, V. Preparation and bioactivity evaluation of bone-like hydroxyapatite nanopowder. *J. Mater. Process. Technol.* **2008**, *202*, 536–542. [CrossRef]

41. Jiang, Y.; Lawrence, M.; Ansell, M.P.; Hussain, A. Cell wall microstructure, pore size distribution and absolute density of hemp shiv. *R. Soc. Open Sci.* **2018**, *5*, 171945. [CrossRef] [PubMed]

42. Stanciu, G.A.; Sandulescu, I.; Savu, B.; Stanciu, S.G.; Paraskevopoulos, K.M.; Chatzistavrou, X.; Kontonasaki, E.; Koidis, P. Investigation of the hydroxyapatite growth on bioactive glass surface. *J. Biomed. Pharm. Eng.* **2007**, *1*, 34–39.

43. Shin, K.; Acri, T.; Geary, S.; Salem, A.K. Biomimetic mineralization of biomaterials using simulated body fluids for bone tissue engineering and regenerative medicine. *Tissue Eng. Part A* **2017**, *23*, 1169–1180. [CrossRef] [PubMed]

44. Radev, L.; Hristov, V.; Michailova, I.; Fernandes, H.M.V.; Salvado, M.I.M. In vitro bioactivity of biphasic calcium phosphate silicate glass-ceramic in $CaO-SiO_2-P_2O_5$ system. *Process. Appl. Ceram.* **2010**, *4*, 15–24. [CrossRef]

45. Algarni, H.; Shahrani, I.; Ibrahim, E.; Eid, R.; Kilany, M.; Ghramh, H.; Ali, A.; Yousef, E. Silver Modified tricalcium phosphate for biomedical application: Structural investigation and study of antimicrobial with histopathological activity. *Sci. Adv. Mater.* **2019**, *11*, 1383–1391. [CrossRef]

46. Zadpoor, A.A. Relationship between in vitro apatite-forming ability measured using simulated body fluid and in vivo bioactivity of biomaterials. *Mater. Sci. Eng. C* **2014**, *35*, 134–143. [CrossRef]

47. Meejoo, S.; Maneeprakorn, W.; Winotai, P. Phase and thermal stability of nanocrystalline hydroxyapatite prepared via microwave heating. *Thermochim. Acta* **2006**, *447*, 115–120. [CrossRef]

48. Destainville, A.; Champion, E.; Bernache-Assollant, D.; Laborde, E. Synthesis, characterization and thermal behavior of apatitic tricalcium phosphate. *Mater. Chem. Phys.* **2003**, *80*, 269–277. [CrossRef]

49. Algarni, H.; Alshahrani, I.; Ibrahim, E.; Eid, R.; Kilany, M.; Ghramh, H.; Ali, A.; Yousef, E. Fabrication and biocompatible characterizations of bio-glasses containing oxyhalides ions. *J. Nanoelectron. Optoelectron.* **2019**, *14*, 328–334. [CrossRef]

50. Clupper, D.C.; Hench, L.L. Bioactive response of Ag-doped tape cast Bioglass®45S5 following heat treatment. *J. Mater. Sci. Mater. Med.* **2001**, *12*, 917–921. [CrossRef]

51. Algarni, H.; AlShahrani, I.; Ibrahim, E.; Eid, R.; Kilany, M.; Ghramh, H.; Abdellahi, M.; Shaaban, E.; Reben, M.; Yousef, E. Synthesis, Mechanical, In Vitro and In vivo bioactivity and preliminary biocompatibility studies of bioglasses. *Sci. Adv. Mater.* **2019**, *11*, 1458–1466. [CrossRef]

MDPI

St. Alban-Anlage 66

4052 Basel

Switzerland

Tel. +41 61 683 77 34

Fax +41 61 302 89 18

www.mdpi.com

Coatings Editorial Office

E-mail: coatings@mdpi.com

www.mdpi.com/journal/coatings